ELEMENTS OF CLASSICAL THERMODYNAMICS

ELEMENTS OF CLASSICAL THERMODYNAMICS FOR ADVANCED STUDENTS OF PHYSICS

BY

A. B. PIPPARD

M.A., Ph.D., F.R.S.

J. H. Plummer Professor of Physics in the University of Cambridge and Fellow of Clare College, Cambridge

CAMBRIDGE

AT THE UNIVERSITY PRESS

1964

PUBLISHED BY THE PRESS SYNDICATE OF THE UNIVERSITY OF CAMBRIDGE
The Pitt Building, Trumpington Street, Cambridge, United Kingdom

CAMBRIDGE UNIVERSITY PRESS
The Edinburgh Building, Cambridge CB2 2RU, UK
40 West 20th Street, New York NY 10011- 4211, USA
477 Williamstown Road, Port Melbourne, VIC 3207, Australia
Ruiz de Alarcón 13, 28014 Madrid, Spain
Dock House, The Waterfront, Cape Town 8001, South Africa

http://www.cambridge.org

First published 1957
Reprinted with corrections 1960
Reprinted 1961
Reprinted with corrections 1964

A catalogue record for this book is available from the British Library

ISBN 0 521 05955 0 hardback
ISBN 0 521 09101 2 paperback

Transferred to digital printing 2004

CONTENTS

PREFACE

The first treatment of thermodynamics which a student of physics gets nowadays is likely to be given in an early year at the university, and to be concerned more with the descriptive aspect of the subject than with its systematic application to a variety of problems. If in a later year he attends a more thorough course it will probably be quite brief, on account of the pressure of other subjects which nowadays rightly rank as of greater urgency, and he cannot normally be expected to supplement his meagre ration of lectures by protracted study of the standard monographs. It is this situation which I have kept in mind in writing a short account of the fundamental ideas of thermodynamics, and to keep it short I have deliberately excluded details of experimental methods and multiplicity of illustrative examples. In consequence this is probably not a suitable text-book for the beginner, but I hope the more advanced student will find here a statement of the aims and techniques, which will illuminate any specialized applications he may meet later. I hope too that for mathematical students whose ambitions point towards theoretical physics it may serve as a concise introduction to the theory of heat. It may be objected by some that I have concentrated too much on the dry bones, and too little on the flesh which clothes them, but I would ask such critics to concede at least that the bones have an austere beauty of their own.

The problems which are provided as an aid to learning are of two kinds. From time to time in the text I have suggested extensions to the arguments which may be profitably developed, and at the end I have placed a small number of exercises, most of which are rather difficult. I am indebted to Mr V. M. Morton for checking these and suggesting improvements.

The relief from teaching which was afforded by an invitation to spend a year as visiting Professor at the Institute for the Study of Metals, in the University of Chicago, provided the stimulus and opportunity for writing this book. I should like to express my gratitude to all at the Institute who made my visit so enjoyable. My warmest thanks are also due to Professor J. W. Stout, Dr D. Shoenberg, F.R.S., and Dr T. E. Faber for their critical comments, and to Mrs Ruth Patterson and Mrs Iris Ross for typing the manuscript.

A. B. P.

May 1957

PREFACE TO FOURTH IMPRESSION

A number of errors and blemishes in the first edition have been pointed out by reviewers and friendly critics, and some of the more serious were corrected in the second printing of that edition. I have taken the opportunity afforded by a further printing to bring some parts of the later chapters up to date, but have otherwise made no substantial changes.

A.B.P.

August 1963

CHAPTER 1

INTRODUCTION

The science of thermodynamics, in the widest sense in which the word is used nowadays, may be said to be concerned with the understanding and interpretation of the properties of matter in so far as they are affected by changes of temperature. In this sense thermodynamics ranks as one of the major subdivisions of physical science, and a variety of mathematical and experimental techniques may be invoked to aid its advancement, with the ultimate aim of providing an explanation of the observed properties of matter at all temperatures in terms of its atomic constitution and the forces exerted by atoms upon one another. This statement covers perhaps a wider field of investigation than can legitimately be called thermodynamics. For example, it can hardly be claimed that the theory of the chemical forces which bind together the atoms of stable chemical compounds is a branch of thermodynamics; rather is it a branch of quantum mechanics in which the concept of temperature plays no part. On the other hand, as soon as we become interested in the excitation or dissociation of molecules as a consequence of heating, the matter becomes truly one in which thermodynamical considerations are involved. In the same way the existence of solids and a great many properties of solids which are only to a minor degree affected by temperature may be explained satisfactorily as purely mechanical consequences of the forces between atoms; thermodynamics strictly enters only when we attempt to account for the temperature-dependent properties, such as heat capacity and (in certain solids) magnetic susceptibility. And, of course, the study of phase transitions (solid to liquid, liquid to gas, or changes of crystalline structure) is in its essence thermodynamical, and provides indeed some of the most interesting problems of present-day thermodynamics.

This is the wide use of the term, but there is a narrower sense in which it is used, and this is what may conveniently be distinguished as *classical thermodynamics*, the subject of this book. Here the method of approach takes no account of the atomic constitution of matter, but seeks rather to derive from certain basic postulates, the laws of thermodynamics, relations between the observed properties of substances. In contrast to the atomic theory of thermal phenomena, classical thermodynamics makes no attempt to provide a mechanistic explanation of why a given substance has the properties observed experimentally; its function is to link together the many observable properties so that they can all be seen to be a consequence of a few. For example, if the equation of state of a gas (the relation between its pressure, volume and temperature) be known, and a determination made of its specific heat at constant pressure over a range of temperatures, then by thermodynamical arguments the specific heat at constant volume may be found, as well as the dependence of both specific heats on pressure or volume. In addition, it may be predicted whether the gas will be heated or cooled when it is expanded through a throttle from a high to a low pressure, and the magnitude of the temperature change may be calculated precisely. This is a comparatively elementary example of the application of classical thermodynamics to a physical problem. The applications discussed in this book should give some idea of the power of thermodynamics to deal with a considerable variety of phenomena, but to appreciate the full scope of the method, and to see how much can be achieved by the use of only simple mathematics, reference should be made to more detailed treatises on the subject. There is none which encompasses the whole field, for the applications of thermodynamics range over many branches of physics, chemistry and engineering, and in each are so extensive as to demand separate treatment if anything like completeness is to be attained. Nevertheless, in spite of a great diversity of methods of presentation, the ideas involved are exactly the same in principle.

The two approaches to thermodynamical theory, the classical or phenomenological approach on the one hand and the statistical approach through molecular dynamics on the other, are so different that it is worth discussing the relationship between them, especially as we shall have no more concern with the latter in what follows. The laws of thermodynamics were arrived at as a consequence of observation and generalization of experience; continued application of the methods of classical thermodynamics to practical problems showed these laws to predict the correct answer in all cases. This is the empirical justification for regarding the laws as having a very wide range of validity. But classical thermodynamics makes no attempt to explain why the laws have their particular form, that is, to exhibit the laws as a necessary consequence of other laws of physics which may be regarded as even more fundamental. This is one of the problems which is treated by statistical thermodynamics. From a consideration of the behaviour of a large assembly of atoms, molecules or other physical entities, it may be shown, with a fair degree of rigour (enough to satisfy most physicists but few pure mathematicians), that those properties of the assembly which are observable by macroscopic measurements are related in obedience with the laws of thermodynamics. This result, which is not derived by considering any very specific model, but which has as wide a range of validity as the laws of mechanics themselves, has perhaps tended to encourage an undervaluation of classical thermodynamics. For from the point of view of the physicist who aims to penetrate as far as possible into the deepest mysteries of the physical world, and to find the fundamental principles from which all physical laws derive, thermodynamics has ceased to be an interesting study, since it is wholly contained within the laws of dynamics.

Not all practitioners, however, of the physical sciences (in which term we may include without prejudice chemists and engineers) have this particular ambition to probe the ultimate mysteries of their craft, and many who have are forced by circumstances to forgo their desire. For these, the great majority, thermodynamics is not so obviously a trivial pursuit; indeed, in many branches it is an almost indispensable tool. For often enough in the pure sciences, and still more in the applied sciences, it is more important to know the relations between the properties of substances than to have a clear understanding of the origin of these properties in terms of the molecular constitution. And even the theoretical physicist who is concerned with a detailed explanation of these properties may find classical thermodynamics a valuable aid, since it reduces the number of problems which require separate statistical treatment—once certain results have been derived the rest follow thermodynamically, as will become clearer in later chapters.

These are some of the reasons which make a study of classical thermodynamics a valuable part of the education of a physical scientist, but there is another, less purely practical reason. The development of the ideas in thermodynamics has a formal elegance which is exceedingly satisfying aesthetically. It has not perhaps quite the rigour of a perfect mathematical proof, but it approaches nearer that logical ideal than almost any other branch of natural science. For this reason alone it may be regarded as an important part of the education of a scientist.

Because classical thermodynamics is capable of so rigorous a formulation it is desirable, in the author's opinion, to present at least the early steps of the argument in a way which brings out the logical development clearly. At the same time a wholly mathematical approach may prove either repellent to the student, or, what is worse, formally intelligible and yet meaningless in terms of physical reality. The early chapters which follow therefore represent some sort of a compromise in which the ideas are expressed in as unmathematical a form as is consistent with exactitude. By the use of more mathematics the arguments could be shortened on paper; it is doubtful whether such a treatment would lead to a speedier assimilation of the ideas by any but the most mathematically minded students.

The pursuit of rigour involves almost inevitably abandoning the historical approach. Great ideas are more often arrived at by a combination of intuition and a judicious disregard of niceties than by a systematic and logical development of explicitly formulated premisses, and certainly the history of thermodynamics bears out this view. But once the goal has been attained it is possible to go back over the road and see how the same end could have been reached more logically. While it is fascinating for the historian of science to see how Carnot, in his astonishing memoir *Sur la puissance motrice du feu* (1824), arrived at so many correct conclusions after having started with the incorrect caloric theory of heat, only confusion would result from trying to base a modern treatment on this work. Carnot's main results were reproduced and extended, principally by James and William Thomson, Clausius and Rankine, after the experiments of Joule (about 1843–9) had provided convincing evidence for interpreting heat as a form of energy and had thus extended the law of conservation of energy to include thermal processes. Even this work, however, suffers from a number of defects from the point of view of logical presentation. For example, it is undesirable in a purely phenomenological development to have recourse to an unobservable atomic interpretation of the nature of heat, if it can be avoided, as it can; and secondly, there is little explicit discussion in this early work of the meaning of that all-important term *temperature*. To be sure, our bodily senses allow us to comprehend with ease the idea of temperature, and without such direct apprehension the development of thermodynamics would surely have been considerably retarded. But that is no reason for continuing to regard the idea of temperature as essentially intuitive, if a satisfactory definition of the term can be given which does not rely on our qualitative sensory impressions only.

We therefore begin our formal treatment of the subject by showing how the ideas of temperature and heat may be systematically formulated on the basis of experiment, so that the subsequent development may be as free as possible from the suspicion that it is based on intuitive concepts or atomistic interpretations.

CHAPTER 2

THE ZEROTH AND FIRST LAWS OF THERMODYNAMICS

Fundamental definitions

Thermodynamics is concerned with real physical systems, which may be solid or fluid, or mixtures of both, or even an evacuated space containing nothing but electromagnetic radiation. Usually the system considered must be contained within a vessel of some kind with which it does not react chemically.

Now the walls of different vessels differ considerably in the ease with which influences from without may be transmitted to the system within. Water within a thin-walled glass flask may have its properties readily changed by holding the flask over a flame or by putting it in a refrigerator; or the change brought about by the flame may be simulated (though not so easily) by directing an intense beam of radiation on to the flask. If, on the other hand, the water is contained within a double-walled vacuum flask with silvered walls (Dewar vessel), the effects of flame or refrigerator or radiation are reduced almost to nothing. The degree of isolation of the contents from external influences can be varied continuously over such a wide range, that it is not a very daring extrapolation to imagine the existence of a vessel having perfectly isolating walls, so that the substance contained within is totally unaffected by any external agency.† Such an ideal wall is termed an *adiabatic wall*, and a substance wholly contained within adiabatic walls is said to be *isolated*. Walls which are not adiabatic are *diathermal walls*. Two physical systems separated from each other only by diathermal walls are said to be in *thermal contact*.

An adiabatic wall may be so nearly realized in practice that it may be claimed to be a matter of experience that when a physical system is entirely enclosed within adiabatic walls it tends towards,

† It is perhaps too much of an extrapolation to imagine the walls impervious to gravitational fields. Rather than postulate the existence of such a wall we shall for the moment avoid problems involving gravitation. It is hoped that when, later in the book, we make occasional reference to gravitational fields the reader will feel enough confidence in his physical understanding of thermodynamics to be able to ignore lacunae in the basic formulation. If he cannot overcome his scruples he must work out a better treatment for himself.

and eventually reaches, a state in which no further change is perceptible, no matter how long one waits. The system is then said to be in *equilibrium*.

Mechanical systems exhibit a number of different types of static equilibrium, which may be exemplified by the behaviour of a spherical ball resting on curves of different shapes and acted upon by gravity. Thermodynamic systems show analogies to some of these, and it is convenient to point them out at this stage, although it is not strictly pertinent to the argument and, indeed, necessitates the use of concepts which have not yet been defined. The following discussion should therefore be regarded as an explanatory parenthesis only. *Stable equilibrium* may be represented by a ball resting at the bottom of a valley; the equilibrium of a pure gas at rest at a uniform temperature is analogous to this. There is no realizable thermodynamic analogue to *unstable equilibrium*,† as of a ball poised at the top of a hill. A ball resting on a flat plane is in *neutral equilibrium*; so is a mixture of water and water vapour enclosed in a cylinder and subjected to a constant pressure by means of a frictionless piston, the whole being maintained at such a temperature that the vapour pressure of the water is exactly equal to the pressure exerted by the piston. For just as the ball may remain at rest at any point on the plane, so the proportion of liquid and vapour may be adjusted at will by movement of the piston. Finally, there is *metastable equilibrium*, represented by a ball resting in a local depression at the top of a hill, and stable with respect to small displacements while unstable with respect to large displacements. It is difficult to find a strict thermodynamic analogue to this type of metastability, but perhaps the nearest approach is exemplified by a supercooled vapour or by a mixture of hydrogen and oxygen. Both the systems have the appearance of stability and may be subjected to small variations of pressure and temperature as if they were truly stable; yet the effect of a condensation nucleus on the former or a spark on the latter shows clearly that they have not the stability of, say, helium gas in equilibrium. The analogy with the ball in metastable equilibrium is not perfect, for these thermodynamic systems are never strictly in equilibrium. Given long enough a supercooled vapour will eventually condense of its own accord, and given long enough a mixture of hydrogen and oxygen will transform itself into water. The time involved may be so enormous, however, perhaps 10^{100} years or more, that the process is not perceptible. For most purposes, provided the rapid change is not artificially stimulated, the systems may be regarded as being in equilibrium.

Although we may discover analogies between thermodynamic and simple mechanical systems it is well to bear in mind an important

† See p. 111 for further discussion of this point.

difference. The equilibrium of a thermodynamic system is never static; the 'matter of experience' mentioned above, that systems tend to a state from which they subsequently do not change, is not strictly an experimental truth. A microscopic examination of minute particles suspended in a fluid reveals them to be in a state of continuous agitation (Brownian movement), and in the same way delicate measurements on a gas would reveal that the density at a given point is subject to incessant minute fluctuations about its mean value. This is, of course, a consequence of the rapid motions of the molecules composing the system, and is an intrinsic property of the system. If one were prepared to wait long enough, and in most cases long enough means a time enormously longer than the age of the universe, one might observe really sizeable departures from the average state of the system. For example, there is no reason why 1 c.c. of a gas, in a state of complete thermal isolation, should not spontaneously contract to half its volume, leaving the other half of the vessel evacuated, and just as suddenly revert to its average state of virtually uniform density. But the whole fluctuation would take only about 10^{-4} sec. to be accomplished, and might be expected to occur once in about $10^{(10^{19})}$ years, so that the possibility of making such an observation need not be seriously contemplated. For most purposes it is quite satisfactory to imagine an isolated system to tend to a definite and invariant state of equilibrium, and classical thermodynamics assumes the equilibrium state to have this static property. In so far as this assumption is not strictly true we must expect to have to revise any results we may derive before applying them to problems in which fluctuations play a significant part. We shall return to this point later (Chapter 7).

Temperature

At this stage it is convenient to consider an especially simple type of thermodynamic system in order to arrive at an idea of the meaning of *temperature*, and we shall for the present confine our attention to homogeneous fluids, either liquids or gases. A gas, of course, of its very nature fills its containing vessel; we imagine any liquid under consideration to be contained exactly by its vessel, leaving no free space for vapour. The especial simplicity of a fluid derives from the fact that its shape is of no consequence thermodynamically; deformation of the containing vessel, if unaccompanied by any change in volume, may in principle be accomplished without the performance of work, and does not alter the thermal properties of the fluid within. By contrast, a solid body can only be altered in shape by the application of considerable stresses, and the thermal properties are in general affected in the process.

It is a fact of experience that a given mass of fluid in equilibrium is completely specified (apart from its shape which is, as just pointed out, of no significance) by a prescription of its volume, V, and pressure, P.† If we take a certain quantity of gas, enclosed in a cylinder with a movable piston, we may fix the volume at some predetermined value and then, with the help of such well-known auxiliary devices as an oven and a refrigerator, set about altering the pressure to any required value. It may readily be verified by experiment that whatever the process by which the pressure and volume are adjusted, the final state of the gas is always the same, no matter what property is examined (e.g. colour, smell, sensation of warmth, thermal conductivity, viscosity, etc.). That is to say, any property capable of quantitative measurement may be expressed as a function of the two variables, P and V.

Let us consider now the behaviour of two systems which are not thermally isolated from one another. If we take two isolated systems and allow them to come into equilibrium separately, and then bring them into thermal contact by replacing the adiabatic wall which divides them by a diathermal wall, we shall find in general that changes take place in both, until eventually the composite system attains a new state of equilibrium, in which the two separate systems are said to be *in equilibrium with one another*. As a matter of sensory experience we know that this is because two systems chosen independently will not in general have the same temperature, and the changes which occur when they are brought into thermal contact result in their eventually attaining the same temperature. But there is no need at this stage to employ the, as yet, meaningless word *temperature* to describe this particular fact of experience. It is sufficient to realize that two systems may be separately in equilibrium and yet not in equilibrium with one another. In particular, two given masses of fluid are not in equilibrium with one another if their pressures and volumes (the *parameters of state*) are chosen arbitrarily. Of the four variables of the composite system, P_1 and V_1 for the first fluid, and P_2 and V_2 for the second, three may be fixed arbitrarily, but for the two fluids to be in equilibrium with one another the fourth variable is then determined by the other three. One may, for example, adjust both P_1 and V_1 by placing the first fluid in an oven until the required values are reached; if then V_2 is fixed it will be found necessary to adjust P_2 by placing the second fluid in the oven before the two fluids are in equilibrium with one another. This may be expressed in a formal manner by saying that for two given masses of fluid there exists a function of the variables of state,

† We leave out of consideration for the moment the possible influences of electric and magnetic fields on the properties of the fluid, assuming such fields to be absent.

$F(P_1, V_1, P_2, V_2)$ such that when the fluids are in equilibrium with one another,

$$F(P_1, V_1, P_2, V_2) = 0. \tag{2·1}$$

The form of the function will depend of course on the fluids considered, and may be determined, if required, by a sufficient number of experiments which measure the conditions under which the fluids are in equilibrium.

In order to establish the existence of the important property *temperature* it is necessary to demonstrate that (2·1) may always be rewritten in the form

$$\phi_1(P_1, V_1) = \phi_2(P_2, V_2), \tag{2·2}$$

in which the variables describing the two systems are separated. This we shall do first by a formal mathematical argument, and then (since to many students the mathematical argument appears too abstract to be altogether meaningful) by an equivalent argument in terms of hypothetical experiments. Both arguments depend on a further fact of experience which has come to be regarded as sufficiently important to be designated the *zeroth law of thermodynamics*:

If, of three bodies, A, B and C, A and B are separately in equilibrium with C, then A and B are in equilibrium with one another.

It is also desirable, for reasons which will appear later, to state what is essentially the *converse of the zeroth law*:

If three or more bodies are in thermal contact, each to each, by means of diathermal walls, and are all in equilibrium together, then any two taken separately are in equilibrium with one another.

The sort of simple experiment upon which the zeroth law is based may be illustrated by the following example. Let C be a mercury-in-glass thermometer, in which the mercury is a fluid at roughly zero pressure (if the thermometer is evacuated) and with a volume determined by its height in the tube; the height of the mercury in a given thermometer is sufficient to determine its state. Then according to the zeroth law if the reading of the thermometer is the same when it is immersed in two different liquids, A and B, nothing will happen when A and B are placed in thermal contact. It is easy enough to multiply examples of the application of this law, which expresses so elementary and common an experience that it was not formulated until long after the first and second laws had been thoroughly established.

Consider now three fluids, A, B and C. The conditions under which A and C are in equilibrium may be expressed by the equation

$$F_1(P_A, V_A, P_C, V_C) = 0,$$

which may be solved for P_C to give an equation of the form

$$P_C = f_1(P_A, V_A, V_C). \qquad (2\cdot3)$$

Similarly, the conditions under which B and C are in equilibrium may be expressed by the equation

$$F_2(P_B, V_B, P_C, V_C) = 0,$$

or, again by solving for P_C,

$$P_C = f_2(P_B, V_B, V_C). \qquad (2\cdot4)$$

Hence the conditions under which A and B are separately in equilibrium with C may be expressed, from (2·3) and (2·4), by the equation

$$f_1(P_A, V_A, V_C) = f_2(P_B, V_B, V_C). \qquad (2\cdot5)$$

But if A and B are separately in equilibrium with C, then according to the zeroth law they are in equilibrium with one another, so that (2·5) must be equivalent to an equation of the form

$$F_3(P_A, V_A, P_B, V_B) = 0. \qquad (2\cdot6)$$

It will be seen now that while (2·5) contains the variable V_C, (2·6) does not. If the two equations are to be equivalent, it can only mean that the functions f_1 and f_2 contain V_C in such a form that it cancels out on the two sides of (2·5) [e.g. $f_1(P_A, V_A, V_C)$ might take the form

$$\phi_1(P_A, V_A)\,\zeta(V_C) + \eta(V_C)].$$

When this cancellation is performed, (2·5) will have the form of (2·2),

$$\phi_1(P_A, V_A) = \phi_2(P_B, V_B),$$

and by an obvious extension of the argument

$$\phi_1(P_A, V_A) = \phi_2(P_B, V_B) = \phi_3(P_C, V_C),$$

and so on for any number of fluids in equilibrium with one another. We have thus demonstrated that for every fluid it is possible to find a function $\phi(P, V)$ of its parameters of state (different of course for each fluid) which has the property that the numerical value of ϕ ($=\theta$, say) is the same for all fluids in equilibrium with one another. The quantity θ is called the *empirical temperature*, and the equation

$$\phi(P, V) = \theta \qquad (2\cdot7)$$

is called the *equation of state* of the fluid. In this way we have shown, by means of the zeroth law, that there exists a function of the state of a fluid, the temperature, which has the property of taking the same value for fluids in equilibrium with one another. Since θ is uniquely determined by P and V,† it is possible to regard the state of the fluid as specified by any two of the three variables P, V and θ.

In order to make the physical meaning of this result somewhat clearer, we shall now show how it may be derived from a consideration of certain simple experiments. Suppose we have two masses of fluid, a standard mass S and a test mass T. Keeping S in a fixed state (P_S and V_S constant) we may vary P_T and V_T of the test mass in such a way as to keep S and T always in equilibrium with one another. Since of the four variables only one is independent (cf. (2·3)), we shall find a relation between P_T and V_T which may be represented by a line such as L in fig. 1. Such a line is termed an *isotherm*. It may now be readily

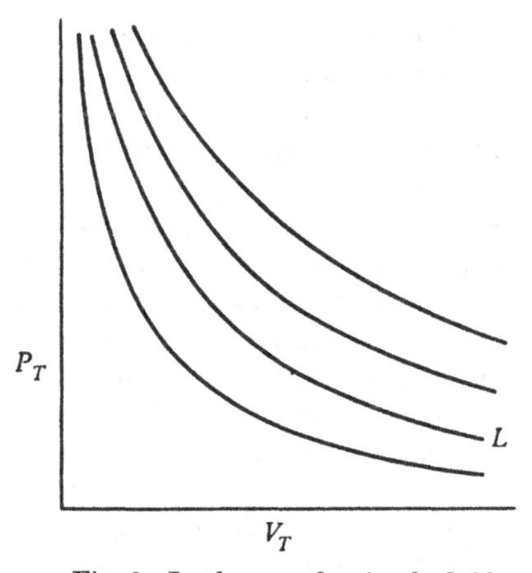

Fig. 1. Isotherms of a simple fluid.

seen that the form of the isotherm is independent of the nature of the standard mass. For suppose we had chosen instead of S another standard mass S' which was in equilibrium with S. Then for any state of T corresponding to a point on L, we should have S in equilibrium with both S' and T; hence, according to the zeroth law, S' and T would also be in equilibrium, and the isotherm of T determined by the use of S' rather than S would pass through this point.

We have shown then how to construct the isotherms of a fluid on the P-V diagram (*indicator diagram*), and also that the isotherms so constructed are determined by the nature of the fluid and not by the choice of the subsidiary standard body. By continued experiment a whole

† Apart from occasional ambiguities arising from multiple solutions of (2·7); e.g. at a pressure of 1 atmosphere water has the same volume at 2° C. and at 6° C.

family of isotherms may be plotted out, as shown schematically in fig. 1. Now let us label each isotherm with a number, θ, chosen at will, which we call the empirical temperature corresponding to the given isotherm. Then provided there is some system, however arbitrary, in the labelling of the isotherms, there will exist a relationship (not necessarily analytic) between P, V and θ which may be written in the same form as (2·7),

$$\phi(P, V) = \theta.$$

Once this labelling of isotherms has been carried out for one particular mass of fluid, however, there exists no latitude of choice so far as other fluids are concerned, if consistency is to be achieved. For the isotherm of a second fluid in equilibrium with the first must be labelled with the same θ. If, and only if, this is done can we say that all fluids having the same value of θ are in equilibrium with one another. This brings us to the same result as was derived before; the two arguments are equivalent.

It is because of the element of choice in the labelling of the isotherms of the first fluid to be selected (the *thermometric body*) that the quantity θ is referred to as the *empirical temperature*. It is usual to choose as the thermometric body a fluid whose properties make a rational choice of θ particularly simple. For example, in a mercury-in-glass thermometer there is effectively only one variable, the volume of the mercury, and θ is taken to be a linear function of the volume. The particular straight line selected depends on the choice of scale; according to the Celsius scale, θ is put equal to 0 at the temperature of melting ice, and 100 at the temperature of water boiling at standard atmospheric pressure. Two fixed points are sufficient to determine the linear relation. Consider now the perfect gas scale of temperature. This is capable of simple definition because of the analytical simplicity of the isotherms, which for perfect gases follow Boyle's law, $PV = \text{constant}$. Thus the equation of state of a perfect gas on any empirical scale must take the form

$$PV = f(\theta),$$

and the nature of the empirical scale determines the form of the function $f(\theta)$. It happens that if the empirical scale is fixed by a mercury-in-glass thermometer, $f(\theta)$ is very nearly a linear function over a wide range of temperature. This experimental result makes it convenient to establish an empirical scale in terms of a perfect gas by adopting as a definition of θ the equation

$$PV = R\theta.$$

The constant R is chosen for any particular mass of gas in such a way that the value of θ shall change by 100 between the melting-point of ice and the boiling-point of water.

For the purpose of the foregoing analysis we have considered only the simplest type of system, a fluid, whose state is definable by two parameters. The argument may easily be extended to more complicated systems, such as solids under the influence of more than simple hydrostatic stresses, or bodies acted upon by electric or magnetic fields, in which cases more than two parameters must be specified in order to determine the state uniquely. The only change involved by this extension is that instead of isothermal lines the body possesses isothermal surfaces in three or more dimensions, and the equation of state may be formally expressed

$$f(x_1, x_2, \ldots, x_n) = \theta,$$

in which $x_1 \ldots x_n$ represent the parameters needed to define the state. The existence of temperature may be proved in exactly the same way as before.

It should be noted that our knowledge of temperature at this stage is insufficient to correlate the empirical temperature of a body with its hotness or coldness. There is no reason why a body having a high value of θ should necessarily be hotter (in the subjective sense, or any other) than one having a low value, since the choice of a temperature scale is entirely arbitrary. It is in fact possible, as in the perfect gas scale, to arrange that the 'degree of hotness' of a body is a monotonic function of its temperature, but we cannot demonstrate this without first inquiring into the meaning of hotness and coldness, and finding a definition which is based on something less subjective than physiological sensation. This involves an investigation of the significance of the term *heat*; only when we have placed this concept on a secure experimental basis can we resolve objectively the relationship between temperature and hotness.

Internal energy and heat

In pursuance of our plan of developing thermodynamics as a phenomenological science, we shall pass over any consideration of the molecular interpretation of heat, and the historical controversies between the followers of the caloric and the kinetic theories. The work of Rumford, Joule and innumerable other experimenters has in truth firmly established the kinetic theory (so firmly, indeed, that it is hard to realize that the matter was extremely controversial little more than a century ago), but although the work itself is of fundamental importance to the phenomenological aspect of thermodynamics, its molecular interpretation is entirely irrelevant.

Let us consider the way in which an experiment, designed to measure the mechanical equivalent of heat, is conducted, taking Joule's paddle-wheel experiment as typical. A mass of water is enclosed in a calorimeter whose walls are made as nearly adiabatic as possible, and through these walls is inserted a spindle carrying paddles, so that mechanical work can be performed on the system consisting of water, calorimeter, paddles and spindle. A measured amount of work, $G\alpha$, is done by applying a known couple, G, to the spindle and rotating it through a known angle, α. As a result of this work the temperature of the water is found to change. The experiment is repeated with a different amount of water in the same calorimeter, and it can then be deduced, after corrections have been applied to allow for the walls being imperfectly adiabatic, what change of temperature a given isolated mass of water would suffer if a given amount of mechanical work were performed on it. An alternative experiment of the same nature may be carried out by replacing the paddle-wheels and spindle by a resistive coil of wire. Work is then performed by passing a measured current, i, through the wire for a measured time, t. If the potential difference across the resistance is E, the work done is Eit (in Joules if practical units are used, in ergs if absolute electromagnetic or electrostatic units). The observed outcome of this experiment is that the same temperature change is produced as in the paddle-wheel experiment by the performance of the same amount of work. In many other similar experiments, using different kinds of mechanical work, the same result is obtained.

It should be most particularly noted that in none of these experiments is any process carried out which can be legitimately called 'adding heat to the system'. All are experiments in which the state of an otherwise isolated mass of water (or other fluid) is changed by the performance of mechanical work. It would be a purely inferential, and phenomenologically quite unjustifiable, interpretation of the experiments to regard the mechanical work as transformed into heat, which then raises the temperature of the water. So long as we take account only of what is observed, the deduction to be drawn from the experiments is one which may be stated in the following generalized form:

If the state of an otherwise isolated system is changed by the performance of work, the amount of work needed depends solely on the change accomplished, and not on the means by which the work is performed, nor on the intermediate stages through which the system passes between its initial and final states.

This statement contains rather more than can be justified by the experimental evidence presented, particularly in its last clause. In order to verify its validity more thoroughly, a different type of experiment is required, in which a given change of state is effected by two different processes, involving markedly different intermediate states of the system. One might, for instance, change the state of an isolated mass of gas from that represented by the point A in fig. 2 to that represented by B, by the two processes indicated, ACB and ADB. In the former the gas is expanded from A to C, and caused to do external work in the process, and is then changed to B at constant volume by means of work supplied, say, electrically. In the latter the electrical work is performed first, to bring the state to D, and the work of expansion second. According to the statement made above the total work performed in each process should be the same. Unfortunately, it does not seem that experiments of this kind have ever been carried out carefully. This is historically merely a consequence of the rapid and universal acceptance of the first law of thermodynamics, and of the kinetic theory of heat, which followed the work of Joule. We must therefore admit that the statement which we have enunciated here, and which is equivalent to the first law of thermodynamics, is not well founded on direct experimental evidence. Its manifold consequences, however, are so well verified in practice that it may be regarded as being established beyond any reasonable doubt.

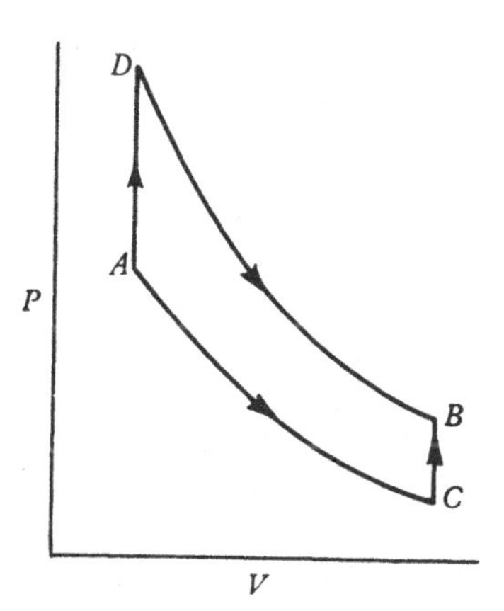

Fig. 2. Different ways of achieving the same change of state.

As a consequence of the first law we may define an important property of a thermodynamic system, its *internal energy*, U. If an isolated system is brought from one state to another by the performance upon it of an amount of work W, the internal energy shall be said to have increased by an amount ΔU which is equal to W. The first law, in stating that W is independent of the path between the initial and final states, ensures that ΔU is determined solely by these states. This means that once U has been fixed arbitrarily at some value U_0 for any one particular state of the system, the value of U for every other state is uniquely determined. It may not be an easy matter in practice to measure the difference in internal energy between any two given states, since the experimental processes involved in getting from one to the other solely by the performance of work on the isolated system may be difficult to accomplish. Nevertheless, however roundabout the journey, a suitable path can always be devised in principle for getting from one given state to another, or vice versa. It will be seen later to be a consequence of the second law of thermodynamics that the path may not necessarily be traced out in both directions under the specified conditions, but for the purpose of determining ΔU it is only necessary to achieve one or the other, not both.

We have therefore shown the existence of a *function of state*, U, that is to say, a function which is determined (apart from an additive constant U_0) by the parameters defining the state of the system. To take a simple fluid as an example,

$$U=U(P,V) \quad \text{or} \quad U=U(P,\theta) \quad \text{or} \quad U=U(V,\theta).$$

For any system, in an *adiathermal* process, i.e. one performed by means of work on an otherwise isolated system,

$$\Delta U=W. \tag{2·8}$$

Up to this point we have considered only systems contained within adiabatic walls, and it is only to these that (2·8) applies. If the system is not so contained, it is found possible to effect a given change of state in different ways which involve different quantities of work. A beaker of water may be brought from 20° to 100° C. by electrical work performed on a resistance immersed in it, or alternatively by lighting a Bunsen burner under it, the latter process involving no work at all. This does not mean that any fault is to be found with the concept of internal energy, but that the equating of ΔU with W is only correct under adiathermal conditions. For any change between given end-states ΔU can always be uniquely determined by carrying out an adiathermal experiment, for which $\Delta U=W$, but if the conditions are not so specialized then in general $\Delta U \neq W$. Instead we may write the equation

$$\Delta U=W+Q, \tag{2·9}$$

and thus define the quantity Q which is a measure of the extent to which the conditions are not adiathermal. The quantity Q so defined is called *heat*, and with the sign convention adopted in (2·9) it is defined to be the heat transferred to the system during the change, just as W is the work done on, rather than by, the system.

Such a manner of introducing and defining heat may appear somewhat arbitrary, and in justification it is necessary to show that the quantity Q exhibits those properties that are habitually associated with heat. In point of fact the properties concerned are not many in number, and may be summarized as follows:

(1) The addition of heat to a body changes its state.

(2) Heat may be conveyed from one body to another by conduction, convection or radiation.

(3) In a calorimetric experiment by the method of mixtures, or any equivalent experiment, heat is conserved if the experimental bodies are adiabatically enclosed.

It hardly requires proof that the quantity Q exhibits properties (1) and (2); from the known existence of diathermal walls it follows that changes of state may be produced not solely by the performance of work, and that the change of state is not necessarily inhibited by the intervention of solid or fluid barriers, or even of evacuated spaces. That Q possesses property (3) is readily demonstrated by consideration of a typical calorimetric experiment, in which two bodies at different temperatures are brought into thermal contact within a vessel composed of adiabatic walls. It is clear, from the definition of U and the fact that the work separately performed on two distinct bodies may be summed to give the total work performed, that U is an additive quantity; if U_1 and U_2 are the internal energies of the two bodies separately, the internal energy of the whole enclosed system, U, may be written as $U_1 + U_2$.

Therefore $\Delta U = \Delta U_1 + \Delta U_2$ in any change in which the bodies maintain their identities. When the bodies are brought into thermal contact, no work is done on either,† and since they are surrounded by adiabatic walls

$$\Delta U = 0,$$

and
$$\Delta U_1 = Q_1, \quad \Delta U_2 = Q_2.$$

Hence
$$Q_1 + Q_2 = 0. \tag{2·10}$$

If Q is interpreted as heat, (2·10) expresses the conservation of heat in the experiment, so that property (3) is demonstrated. We see then that Q possesses all the properties habitually associated with heat, and the use of the term is justified. Moreover, we may now understand the significance, as expressed by (2·9), of the common statement of the *first law of thermodynamics*:

Energy is conserved if heat is taken into account.

Let us now return to the question of what is meant by the terms *hotter* and *colder*, which, as we have seen earlier, do not necessarily bear any relation to higher and lower temperatures on an empirical scale. In the experiment just analysed, we found that the gain of heat by one body equalled the loss by the other, and this behaviour, although it does not necessarily imply that heat is a physical entity whose movement can be followed from one body to another, is called, for the sake of convenience, the *transfer of heat* from one body to the other. In general if any two bodies are brought into thermal contact under such conditions that no work is performed on either, there will be a transfer of heat with accompanying change of state of both, unless they are at the

† We are neglecting the work done by the surrounding atmosphere if the volumes of the bodies alter during the experiment. The argument may readily be extended to include this effect, without modification of the conclusion.

same temperature and consequently in equilibrium with one another. The rate at which the transfer occurs may usually be varied over a wide range by altering the nature of the diathermal wall separating the bodies. The rate is a measure of the *thermal conductance* of the wall. The body which loses heat (Q negative) is said to be *hotter* than that which gains heat (Q positive), and the latter is said to be *colder*.

It now remains to demonstrate that the scale of hotness, so defined, may be consistently linked with a scale of temperature, in the sense that all bodies at a temperature θ_1 shall be hotter than all bodies at a temperature θ_2, if θ_1 is greater than θ_2. To prove that this is possible we consider the consequences of presuming it to be impossible. Given three bodies, A, B and C, of which A is at a temperature θ_1 and B and C at a temperature θ_2, let us suppose that A is hotter than B, and that C is hotter than A. We may now vary the temperature of B slightly so that it is hotter than C while still colder than A. If the three bodies are then placed in thermal contact, heat will be transferred from A to B, from B to C, and from C to A; by adjustment of the thermal conductances the rate of transfer may be made the same at each contact, and a state of (dynamic) equilibrium will be established. But if any two of the bodies are taken away they will be found not to be in equilibrium, since all are at different temperatures, and this is in conflict with the converse to the zeroth law (p. 9). We conclude then that if a body at θ_1 is hotter than any one body at θ_2 it is hotter than all bodies at θ_2. Therefore if we take a suitable thermometric substance and label its isotherms in such a way that successively hotter isotherms are ascribed successively higher empirical temperatures, we have achieved a correlation between temperature and hotness which is equally valid for all substances. Henceforth we shall take the term 'higher temperature' to imply 'of greater degree of hotness'.

In the above argument it has been assumed that no performance of work accompanied the transfer of heat from one body to another. This is unnecessarily restrictive; all that is necessary is that such parameters as are needed, together with the temperature, to define the states of the bodies shall remain constant when the thermal contact is made. The correlation of temperature and hotness then follows exactly as above. This has the important consequence that addition of heat to a body whose independent parameters of state, apart from the temperature, are maintained constant always causes a rise of temperature, and therefore the principal specific heats† of a body are always positive.

† Here and elsewhere we use the term *specific heat* to mean the thermal capacity of the thermodynamic system, regardless of its size or nature. If the specific heat is measured under such conditions that the independent parameters of state (other than the temperature) are mantained constant, the measured property is one of the *principal specific heats* (e.g. C_P and C_V for a fluid).

CHAPTER 3

REVERSIBLE CHANGES

Reversibility and irreversibility

In this chapter we shall discuss the significance and some applications of the first law of thermodynamics:

$$\Delta U = Q + W \quad \text{in any change.}$$

However satisfactory this equation may be as an expression of a physical law, from an analytical point of view it leaves much to be desired. For although U is a well-behaved function of state in the sense that ΔU depends only on the initial and final states of the system and not on the path by which the change is effected, the same does not apply to either Q or W. There do not exist functions of state, of which Q and W are finite differences, as may be illustrated by a simple example. The temperature of a vessel of water may be raised either by performing work (as in Joule's experiments) or by supplying heat; thus for any given change the values of Q and W may be altered at will, only their sum remaining constant. And this is true not only for finite, but also for infinitesimal changes, which we may represent by the equation

$$\mathrm{d}U = q + w. \tag{3·1}$$

It is quite impermissible to write the increments q and w in the form $\mathrm{d}\mathscr{Q}$ and $\mathrm{d}\mathscr{W}$, and to regard them as differential coefficients of hypothetical functions of state, $\mathscr{Q}$ and $\mathscr{W}$, since there is no sense in which a given body can be said to contain a certain total quantity of heat, $\mathscr{Q}$, or a certain total quantity of work, $\mathscr{W}$. All that can be said along these lines is that the internal energy U is well defined (apart from a certain arbitrariness in fixing the zero of energy), and that U may be altered by means of work or of heat, the two contributions being usually subject to some measure of arbitrary variation, according to the method by which the change is effected.

Nevertheless, if certain restrictions are imposed on the way in which the change occurs, it is possible to treat q and w as well-behaved differential coefficients, $\mathrm{d}Q$ and $\mathrm{d}W$, without becoming involved in mathematical absurdities. For if the conditions of an infinitesimal change are such that either q or w depends only on the initial and final states and not on intermediate states, it follows, since $\mathrm{d}U$ has this property, that the third term in the equation must also be independent of path. Then (3·1) may legitimately be written in the form

$$dU = dQ + dW, \tag{3·2}$$

and each term will have a unique value for any given infinitesimal change.

We must now inquire what are the restrictive conditions which validate this procedure. Clearly one simple possibility is to carry out the change under such conditions that either q or w vanishes; then (3·1) takes either of the forms $dU = w$ or $dU = q$, and clearly w and q are uniquely defined and may be treated as differential coefficients, dW and dQ. A more interesting state of affairs comes about, however, if neither q nor w vanishes, but w is expressible as a function of the parameters of state. This may conveniently be illustrated by reference to a simple fluid. If the fluid is contained within a cylinder, and volume changes are produced by movement of a piston, the work done on the system (fluid + cylinder and piston) in an infinitesimal volume change dV is $-P'dV$, where P' is the pressure exerted by an outside agency on the piston. In particular, if the piston moves without friction the pressure P' must be the same as the pressure P in the fluid if the system is to remain in equilibrium, and the work w in a small change of volume is just $-PdV$. Now for any infinitesimal change between given states of the fluid, the value of P is uniquely defined by the state, as is also the value of dV by the change considered, so that w takes a definite value, and we may write (3·2) in the form

$$dU = dQ - PdV. \tag{3·3}$$

Note, however, that care must still be exercised when finite changes are considered, for in such a change the work done is $-\int PdV$, and the value of this integral, unlike that of $\int dU$, depends on the path of integration. So that although dQ is a well-defined quantity for any infinitesimal change, it is not an exact differential, that is, the differential coefficient of a function of state.

In order that we may equate w to $-PdV$ in this case, two conditions must be satisfied, first that the change must be performed slowly, and secondly that there must be no friction. It is easy to understand the reason for these conditions if we imagine experiments in which they are violated. Suppose the fluid to be a gas, and the piston to be withdrawn a small distance suddenly; then a rarefaction of the gas will be produced just inside the piston, and the work done by the gas will be less than if the pressure were uniform. An extreme example of this behaviour occurs in Joule's experiment, shown schematically in fig. 3.

Gas is contained at uniform pressure within a vessel, of which one wall separates it from an evacuated chamber. If this wall is pierced the gas expands to fill the whole space available, but it does so without performing any external work; w is zero even though $P\,\mathrm{d}V$ does not vanish. If we are to equate w to $-P\,\mathrm{d}V$ it is essential to perform the expansion so slowly that the term $-P\,\mathrm{d}V$ has a meaning. During the course of a rapid expansion the gas becomes non-uniform, so that there is no unique pressure P. In fact the expansion must take place sufficiently slowly that the gas is changed from its initial to its final state by the process of passing through all intermediate states of equilibrium, so that at any time its state may be represented by a point on the indicator diagram. Such a change is called *quasistatic*. In the experiment shown in fig. 3 only the initial and final states fall on the indicator diagram; any intermediate state would need for its adequate representation a specification of the distribution of pressure and velocity at all points in the vessel—requiring obviously many more than the two parameters of state which are sufficient for a fluid in equilibrium. Strictly, of course, if the fluid is to remain in equilibrium at all times during a change of state the change must be performed infinitely slowly. Usually in practice, however, a fluid requires very little time to equalize its pressure, and the change can be carried out quite quickly without provoking any significant departure from equilibrium.

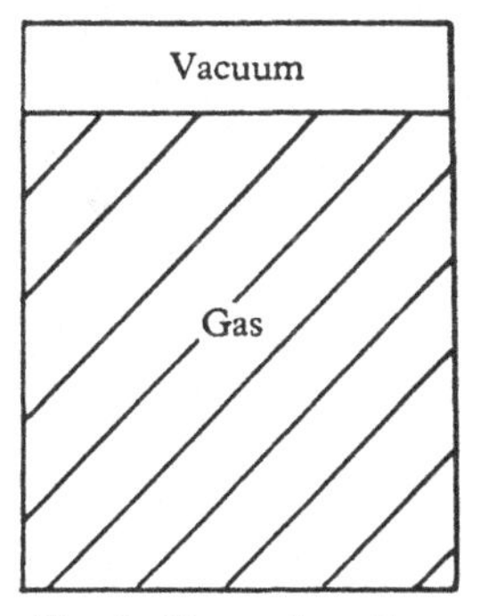

Fig. 3. Expansion of gas into a vacuum.

Consider now the effect of friction in the expansion or compression of a gas. When the gas is being compressed the pressure, P', to be exerted by the external agency must exceed the pressure, P, of the gas itself if there is friction between the cylinder and piston, and when the gas is being expanded P' must be less than P. In a cycle of compression and expansion, then, the relation between P' and V is as illustrated in fig. 4, while the broken line shows the relation between P and V. Now in any infinitesimal change of volume the work done by the external agency is $-P'\,\mathrm{d}V$, and clearly for a change between given initial and final states, the value of this quantity depends on the magnitude of the frictional forces and on the direction in which the change is carried out. Indeed, during the course of the cycle, at the end of which the gas has been returned to its initial state, the work done by the external agency is $-\oint P'\,\mathrm{d}V$, which is positive and equal to the area of the cycle in fig. 4. Since the internal energy of the gas has been returned to its initial value, the cycle shown can only be performed if an amount of heat $-\oint P' \mathrm{d} V$ is extracted from the system during the cycle. We may say then that in this process work has been converted into heat through the effect of the frictional forces. By this we do not intend to imply any particular microscopic mechanism, although an analysis of the frictional process from an atomic point of view would surely reveal the degradation of ordered mechanical energy into disorderly motion of the atoms composing the system. From the point of view of classical thermodynamics this is merely a special example of a very common type of process which we call for convenience *conversion of*

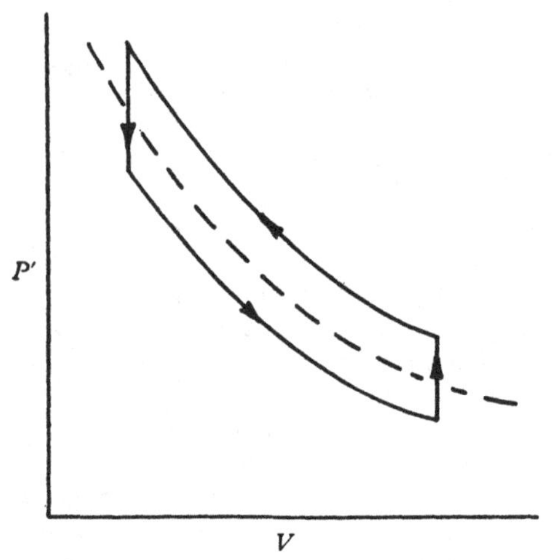

Fig. 4. Effect of friction on P-V relation.

work into heat, and which may be generally defined as the use of mechanical work to perform a process which could be equally well performed by means of heat. All determinations of the mechanical equivalent of heat are examples of the conversion of work into heat in this sense.

This, however, is somewhat of a digression from our main argument, which is that friction prevents the equating of w to $-P \mathrm{d} V$ and hence prevents both w and q in (3·1) from being treated as well-behaved differential coefficients. This analysis of a special example should enable us to see the importance of that particular class of idealized processes which are designated *reversible*. A *reversible process* is defined as one which may be exactly reversed by an infinitesimal change in the external conditions. If then a fluid is to be compressed reversibly there must be no friction between piston and cylinder, otherwise the pressure applied externally to the piston would need to be reduced by a finite amount before the gas could be expanded. In the same way it may be seen that compressions and expansions must be performed slowly, in order to avoid non-uniformities of pressure in the fluid which would prevent the process from being reversible in the sense defined above. In fact, the condition of reversibility in a change is such as to ensure first that the system passes from one state to another without appreciably deviating from a state of equilibrium, and secondly that the external forces are uniquely related to the internal forces. It should also be noted, though the significance of this will not appear until later, that in order to transfer heat reversibly from one body to another the hotter of the two bodies should be only infinitesimally hotter, so that by an infinitesimal reduction of its temperature it may become the cooler, and the direction of heat transfer may be reversed.

To sum up, if a fluid be caused to undergo an infinitesimal change reversibly, it is legitimate to apply the first law to the change in the form (3·3),

$$\mathrm{d}U = \mathrm{d}Q - P\,\mathrm{d}V.$$

Different types of work

So far we have only considered in any detail the reversible performance of work by means of pressure acting on a fluid. Let us now see how (3·3) may be extended so as to apply to situations in which other types of work are involved. There are a number of simple extensions which may be written down immediately, such as the work done by a tensile force in stretching a wire, or the work done in enlarging a film against the forces of surface tension. In the former case, if a wire is held in tension by a force f and increases in length by $\mathrm{d}l$, the work done by the external agency exerting the force is $f\,\mathrm{d}l$. Provided that the wire is deformed elastically, and not plastically, the change of length will be reversible, and we may apply the first law to the wire in the form

$$\mathrm{d}U = \mathrm{d}Q + f\,\mathrm{d}l. \tag{3·4}$$

It should be noted that the validity of this equation does not depend on the wire obeying Hooke's law, but only on the absence of elastic hysteresis so that at a constant temperature stress and strain are uniquely related. The corresponding result for a film having surface tension γ and area $\mathscr{A}$ takes a similar form

$$\mathrm{d}U = \mathrm{d}Q + \gamma\,\mathrm{d}\mathscr{A}, \tag{3·5}$$

subject again to the condition that γ shall be a unique function of $\mathscr{A}$ at constant temperature. This condition is normally satisfied by γ being practically independent of $\mathscr{A}$.

Somewhat more careful consideration is needed for the important extension of the argument to situations where a body is influenced by electric and magnetic fields, and we shall consider the latter case in some detail. There has been in the past a certain amount of confusion concerning the correct formulation of the expression for work performed by means of a magnetic field, but no difficulty need arise if we specify exactly what the experimental arrangement is which we are seeking to analyse. This is, in any case, a highly proper attitude, for thermodynamics is above all a systematic formulation of the results of experiment, and we need feel no shame at keeping in mind a concrete physical example of the problem under consideration. Let us therefore suppose that we produce a magnetic field by means of a solenoid which is excited by a battery, as in fig. 5. In practice the solenoid would normally possess electrical resistance, but as the result we shall obtain is found by more detailed analysis to be independent of the resistance, we may simplify the argument by taking it to be zero. Then the function of the battery is to provide the work needed to create or change the field in the solenoid. It may be regarded as being capable of giving an e.m.f. which is variable at will; and so long as conditions in the solenoid remain steady its e.m.f. may be reduced to zero. The thermodynamic system is now taken to consist of everything within the enclosure marked by broken lines in fig. 6, the solenoid and magnetizable body. Heat may be transmitted through the walls of the enclosure direct to the body (we may neglect any thermal effects in the solenoid itself), and work may be done on the system by the battery.

Consider first the empty solenoid (fig. 5). If at any point the field is $\mathscr{H}$ the magnetic energy is known from elementary electromagnetic considerations to be given by $\frac{1}{8\pi}\int_{\text{all space}}\mathscr{H}^2\,\mathrm{d}V$.† This is the work which the battery must do in creating the field. We shall not assume $\mathscr{H}$ to be constant within the solenoid, but shall suppose that at every point it is proportional to the current, i, through the battery,

$$\mathscr{H}=\mathbf{h}i, \tag{3·6}$$

where $\mathbf{h}$ is a vector function of position. Now consider the effect produced by the body. If its magnetization changes for any reason, a back e.m.f. will be induced in the solenoid, and the battery will have to provide an e.m.f. of the same magnitude in order to maintain the current constant; similarly, a back e.m.f. will be produced if $\mathscr{H}$ is varied by a change of i. The rate at which work is being done by the battery on the system is just this e.m.f. times i, and, for example, if the field of the empty solenoid is changed,

$$\frac{\mathrm{d}W}{\mathrm{d}t}=\frac{\mathrm{d}}{\mathrm{d}t}\left\{\frac{1}{8\pi}\int\mathscr{H}^2\,\mathrm{d}V\right\}. \tag{3·7}$$

† All magnetic terms are expressed in electromagnetic units.

In order to calculate the rate of working due to changes in the magnetization of the body, we make use of two simple principles: first, that the law of superposition holds, so that each element of the body may be treated independently of the rest, and the total effect found by integration over the whole body; and secondly, that the back e.m.f. induced by the creation of an elementary magnetic dipole does not depend on the nature of the dipole, but only on the rate of change of the dipole moment and a geometric factor depending on the solenoid. We may therefore calculate the work done in creating an elementary dipole $\mathbf{dm}$ by use of any simple model dipole, and for our purpose a small current loop, as in fig. 7, is as convenient as any. Suppose then

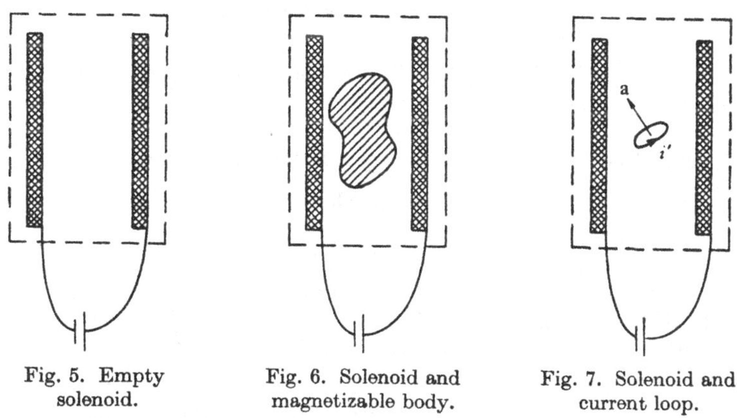

Fig. 5. Empty solenoid.

Fig. 6. Solenoid and magnetizable body.

Fig. 7. Solenoid and current loop.

Figs. 5–7. Calculation of work in reversible magnetization.

we have a loop of area $\mathbf{a}$ (the vector being directed normal to the plane of the loop) carrying a current i'. Since unit current in the solenoid produces a field $\mathbf{h}$ at the loop, the flux linkage due to unit current is $\mathbf{h}.\mathbf{a}$, and this is the mutual inductance between the loop and the solenoid. If then i' changes, the e.m.f. generated in the solenoid is $\mathbf{h}.\mathbf{a}\,\mathrm{d}i'/\mathrm{d}t$, so that the battery must work at a rate $i\mathbf{h}.\mathbf{a}\,\mathrm{d}i'/\mathrm{d}t$. But by Ampère's theorem $\mathbf{a}i'$ is the magnetic moment of the loop, and by (3·6) the solenoid field at the loop is $\mathbf{h}i$, so that we may write for this elementary process,

$$\frac{\mathrm{d}W}{\mathrm{d}t}=\mathscr{H}.\frac{\mathrm{d}\mathbf{m}}{\mathrm{d}t}. \tag{3·8}$$

Now, removing the time derivatives in (3·7) and (3·8), and integrating over all space, we have the equation

$$\mathrm{d}W=\mathrm{d}\left\{\frac{1}{8\pi}\int\mathscr{H}^2\,\mathrm{d}V\right\}+\int(\mathscr{H}.\mathrm{d}\mathscr{I})\,\mathrm{d}V, \tag{3·9}$$

in which $\mathscr{I}$ is the intensity of magnetization (the magnetic moment $\mathrm{d}\mathfrak{m}$ of an element $\mathrm{d}V$ is $\mathscr{I}\mathrm{d}V$). This may also be written in the form

$$\mathrm{d}W = \frac{1}{4\pi}\int(\mathscr{H}.\mathrm{d}\mathscr{B}')\,\mathrm{d}V, \tag{3·10}$$

in which the new vector $\mathscr{B}'$ is defined as $\mathscr{H}+4\pi\mathscr{I}$. It should be particularly noted that this is not the same as the conventional vector $\mathscr{B}$ (*induction*), for the field variable $\mathscr{H}$ which enters into $\mathscr{B}'$ is not the local value of the magnetic field within the body, but the field due to the solenoid in the absence of the body; it is in fact what is usually called the *external* or *applied field*.†

If the body is situated in a uniform external field, (3·9) takes the form,

$$\mathrm{d}W = \mathrm{d}\left\{\frac{1}{8\pi}\int\mathscr{H}^2\mathrm{d}V\right\} + \mathscr{H}.\mathrm{d}\mathscr{M}, \tag{3·11}$$

in which $\mathscr{M}$ is written for $\int\mathscr{I}\mathrm{d}V$, the magnetic moment of the whole body. This is the form which is applicable to most problems of interest.

The results which we have derived express quite generally the work done in the particular type of experiment envisaged. We may, however, only substitute any of these expressions for $\mathrm{d}W$ in (3·2) and regard $\mathrm{d}W$ as a well-behaved differential coefficient if the change is reversible. For this reason we must exclude ferromagnetic and other substances which exhibit hysteresis from our future discussion, just as we had to exclude plastic deformation in considering the stretching of a wire. Considerable attention has been paid to the question of how to include hysteresis effects in the framework of classical thermodynamics, but it is doubtful whether any really satisfactory treatment of such problems can be given without a more detailed microscopic treatment, which falls outside our scope. We shall therefore only consider applications to substances for which $\mathscr{I}$ is uniquely (not necessarily linearly) related to the local field strength, and to which the first law can be written in either of the forms

$$\mathrm{d}U = \mathrm{d}Q + \mathrm{d}\left\{\frac{1}{8\pi}\int\mathscr{H}^2\mathrm{d}V\right\} + \mathscr{H}.\mathrm{d}\mathscr{M} + \text{other reversible work terms}, \tag{3·12}$$

or

$$\mathrm{d}U' = \mathrm{d}Q + \mathscr{H}.\mathrm{d}\mathscr{M} + \text{other reversible work terms}. \tag{3·13}$$

† It is also possible to express $\mathrm{d}W$ as $\frac{1}{4\pi}\int(\mathscr{H}_l.\mathrm{d}\mathscr{B}_l)\mathrm{d}V$, in which $\mathscr{H}_l$ and $\mathscr{B}_l$ are the local values of the magnetic field and induction, but the form (3·10) is usually more convenient in practice. A proof that these forms are equivalent is given by V. Heine, *Proc. Camb. Phil. Soc.* **52**, 546 (1956).

In (3·13) the term expressing the field energy of the empty solenoid has been incorporated in U', which equals $U-\frac{1}{8\pi}\int\mathscr{H}^2\,\mathrm{d}V$.

The corresponding case of electric fields and dielectric bodies may be treated in a quite analogous manner, taking, for example, a condenser as the source of the electric field. Then it is found that, expressed in the electrostatic system of units,

$$\mathrm{d}W=\mathrm{d}\left\{\frac{1}{8\pi}\int\mathscr{E}^2\,\mathrm{d}V\right\}+\int(\mathscr{E}\,.\,\mathrm{d}\mathscr{P})\,\mathrm{d}V, \tag{3·14}$$

in which $\mathscr{E}$ is the electric field in the empty condenser, and $\mathscr{P}$ is the dipole moment per unit volume. Equation (3·14) is analogous to (3·9), and the analogues of (3·10–3·13) may easily be constructed.

The results we have derived are typical of the form which $\mathrm{d}W$ takes in a reversible change, and can be represented in general by the product of a generalized force X and a generalized infinitesimal displacement $\mathrm{d}x$, so that the first law may be written for reversible changes:

$$\mathrm{d}U=\mathrm{d}Q+\sum_i X_i\,\mathrm{d}x_i. \tag{3·15}$$

In the examples considered, $X\,\mathrm{d}x$ takes the forms $-P\,\mathrm{d}V, f\mathrm{d}l, \gamma\,\mathrm{d}A$, $\frac{1}{4\pi}\int(\mathscr{H}\,.\,\mathrm{d}\mathscr{B}')\,\mathrm{d}V$† and $\frac{1}{4\pi}\int(\mathscr{E}\,.\,\mathrm{d}\mathscr{D}')\,\mathrm{d}V$; similar forms may be constructed for changes of state which involve other physical variables. In many practical applications of thermodynamics there is only one work term of importance. We shall, for the sake of illustration, usually consider the simple fluid as typical of these applications and write $\mathrm{d}W$ as $-P\,\mathrm{d}V$, but it should be remembered that all the results so derived may be transformed to meet different situations by changing P and V into different forms of X and x (care being taken, of course, to alter the signs and supply numerical coefficients where necessary).

By restricting ourselves to reversible changes we have been enabled to write the first law in the form of a differential equation (3·2) in which, however, only one term is an exact differential. But our analysis of various physical systems has shown how the inexact differential $\mathrm{d}W$ may be written in the form $\sum_i X_i\,\mathrm{d}x_i$, in which all the X_i and x_i are functions of state, and we have thereby advanced considerably towards casting the law into a form in which it may be manipulated mathematically with ease. Nevertheless, we have the inexact differential $\mathrm{d}Q$ still not expressed in terms of functions of state. The problem of

† The reader should consider carefully what are the analogues of X and x in this expression.

finding the appropriate form for dQ is one which cannot be solved by means of the first law alone; new experimental evidence must be invoked, and this evidence is summed up in the second law of thermodynamics. With its aid we may do for dQ what we have in this chapter done for dW, and at the same time remove the restriction on the application of our results solely to reversible changes.

CHAPTER 4

THE SECOND LAW OF THERMODYNAMICS

The first law of thermodynamics expresses a generalization of the law of conservation of energy to include heat, and thereby imposes a formidable restriction on the changes of state which a system may undergo, only those being permitted which conserve energy. However, out of all conceivable changes which satisfy this law there are many which do not occur in practice. We have already implicitly noted the fact that there is a certain tendency for changes to occur preferentially in one direction rather than for either direction to be equally probable. For example, we have taken it as a basic assumption, in accord with observation, that systems left to themselves tend towards a well-defined state of equilibrium. It is not observed that a reversion to the original non-equilibrium state occurs; indeed, if it did it would be very doubtful whether the term *equilibrium* would have any meaning. Again we have found it valuable to make a distinction between reversible and irreversible changes; the former are as readily accomplished in the backward as in the forward direction, the latter have not this property. It is the irreversible change which should be regarded as the normal type of behaviour. In order that a change shall occur reversibly very stringent conditions must be imposed which clearly make it a limiting case of an irreversible change. Strictly speaking the reversible change is an abstract idealization—all changes which occur in nature are more or less irreversible, and exhibit therefore a preferential tendency. However, the idea that there is a preferred direction for a given change has been perhaps most clearly expressed in our discussion of the terms *hotter* and *colder*. There is an unmistakable tendency for heat to flow from a body of higher temperature to one of lower temperature rather than for either direction of flow to occur spontaneously. The second law of thermodynamics is little more than a generalization of these elementary observations. In essence it states that there is no process devisable whereby the natural tendency of heat to flow from higher to lower temperatures may be systematically reversed. Of the many statements of the law which have been proposed, that of Clausius approaches most clearly this point of view:

It is impossible to devise an engine which, working in a cycle, shall produce no effect other than the transfer of heat from a colder to a hotter body.

The clause 'working in a cycle' should be noted. Only by the use of a cyclical process can it be guaranteed that the process is exactly repeatable so that the amount of heat which can be transferred is unlimited. It is easy to devise non-cyclical processes which transfer heat from a colder to a hotter body. Consider, for instance, a quantity of gas contained in a cylinder and in thermal contact with a cold body. The gas may be expanded to extract heat from the body; if it is then isolated and compressed it will become hotter, and may be brought into equilibrium with a hot body; further compression will enable heat to be transferred to the hot body. But at the end of this process the gas is not in its original state, and no violation of the second law has occurred yet, even though the total amount of work done by or on the gas may have been made to vanish. Only if the gas can be brought back to its original state without undoing the heat transfer already effected can any violation of the second law be claimed. In fact, no violation can be brought about in this case, nor with any of the ingenious and often subtle engines which have been devised with the object of circumventing the law. Moreover, the consequences of the law are so unfailingly verified by experiment that it has come to be regarded as among the most firmly established of all the laws of nature.

Kelvin's formulation of the second law is very similar to that of Clausius, with its emphasis rather more on the practical engineering aspect of heat engines:

It is impossible to devise an engine which, working in a cycle, shall produce no effect other than the extraction of heat from a reservoir and the performance of an equal amount of mechanical work.

Kelvin's law denies the possibility of constructing, for example, an engine which takes heat from the atmosphere and does useful work, while at the same time giving out liquid air as a 'waste product'. It is not difficult to show that a violation of Clausius's law enables Kelvin's law to be violated, and vice versa, so that the two laws are entirely equivalent; a proof of this assertion is left as an exercise for the reader.

A third formulation, due to Carathéodory, is not so clearly related:

In the neighbourhood of any equilibrium state of a system there are states which are inaccessible by an adiathermal process.

Each formulation has its enthusiastic supporters. For the engineer or the practically minded physicist Clausius's or Kelvin's formulations are more directly meaningful, and, moreover, the derivation from them of the important consequences of the law may be made without a great deal of mathematics. On the other hand, Carathéodory's formulation is undoubtedly more economical, in that it demands the impossibility of a rather simpler type of process than that considered in the other formulations. A typically impossible process is the cooling of the water in Joule's paddle-wheel experiment; this is a very simple example since there is only one means whereby work may be performed, and the impossibility of cooling arises from the obvious fact that the paddle-wheel can only do work on the water and not extract energy. But Carathéodory asserts that however complex the system, however many the different forms of work involved, which may be both positive and negative, it is still true that not all changes may be accomplished adiathermally. Carathéodory's law appears to differ in outlook from the others. The average physicist is prepared to take Clausius's and Kelvin's laws as reasonable generalizations of common experience, but Carathéodory's law (at any rate in the author's opinion) is not immediately acceptable except in the trivial cases, of which Joule's experiment is one; it is neither intuitively obvious nor supported by a mass of experimental evidence. It may be argued therefore that the further development of thermodynamics should not be made to rest on this basis, but that Carathéodory's law should be regarded, in view of the fact that it leads to the same conclusions as the others, as a statement of the minimal postulate which is needed in order to achieve the desired end. It bears somewhat the same relation to the other statements as Hamilton's principle bears to Newton's laws of motion.

In view of the wealth of discussion which has centred on the development of the consequences of the second law, no harm will arise from giving several different approaches, not always in complete detail. The reader may then select which he prefers, or, if none is to his taste, consult other texts for further varieties of what is basically the same argument. In everything that follows we shall be closely concerned with *adiabatic* changes, that is, reversible adiathermal changes† which may be analytically represented by putting $\mathrm{d}Q$ equal to zero in (3·15):

$$\mathrm{d}U - \sum_i X_i \mathrm{d}x_i = 0 \quad \text{for an adiabatic change.} \tag{4·1}$$

If U is treated as a function of the generalized position coordinates x_i and of one other coordinate X_r then

$$\mathrm{d}U = \sum_i \frac{\partial U}{\partial x_i}\mathrm{d}x_i + \frac{\partial U}{\partial X_r}\mathrm{d}X_r,$$

so that an adiabatic change may be represented by the equation

$$\sum_i \left(\frac{\partial U}{\partial x_i} - X_i\right)\mathrm{d}x_i + \frac{\partial U}{\partial X_r}\mathrm{d}X_r = 0, \tag{4·2}$$

† Some writers prefer to use *adiabatic* in the sense of our *adiathermal*, and to call our *adiabatic* change a *reversible adiabatic* or *isentropic* change.

which for convenience we may rewrite in the form

$$\sum_i Y_i \mathrm{d}y_i = 0, \tag{4·3}$$

in which all the coordinates x_i and X_r are relabelled y_i, and each coefficient Y_i is a function of state.

For a simple fluid, (4·3) takes the form

$$\left\{\left(\frac{\partial U}{\partial V}\right)_P + P\right\}\mathrm{d}V + \left(\frac{\partial U}{\partial P}\right)_V \mathrm{d}P = 0$$

or

$$\left.\begin{aligned} &\mathrm{d}P/\mathrm{d}V = F(P, V), \\ \text{where}\quad &F(P, V) = -\left\{\left(\frac{\partial U}{\partial V}\right)_P + P\right\}\Big/\left(\frac{\partial U}{\partial P}\right)_V. \end{aligned}\right\} \tag{4·4}$$

At every point on the indicator diagram of a fluid, the gradient of the adiabatic is uniquely defined by (4·4), and a step-by-step integration of the equation, starting from any point, will lead to a unique adiabatic line. Thus (4·4) represents a family of adiabatic lines covering the whole indicator diagram.

As soon, however, as we go to more elaborate systems, having more than two independent parameters of state, the situation alters. A three-parameter adiabatic equation,

$$Y_1 \mathrm{d}y_1 + Y_2 \mathrm{d}y_2 + Y_3 \mathrm{d}y_3 = 0, \tag{4·5}$$

does not necessarily represent a family of surfaces in three-dimensional space. Only, in fact, if the coefficients in (4·5) satisfy the equation

$$Y_1\left(\frac{\partial Y_2}{\partial y_3} - \frac{\partial Y_3}{\partial y_2}\right) + Y_2\left(\frac{\partial Y_3}{\partial y_1} - \frac{\partial Y_1}{\partial y_3}\right) + Y_3\left(\frac{\partial Y_1}{\partial y_2} - \frac{\partial Y_2}{\partial y_1}\right) = 0, \tag{4·6}$$

can (4·5) be integrated as a family of surfaces. For example, the equation

$$y_1 \mathrm{d}y_1 + y_2 \mathrm{d}y_2 + y_3 \mathrm{d}y_3 = 0 \tag{4·7}$$

may be integrated immediately as the family of spheres

$$y_1^2 + y_2^2 + y_3^2 = \text{constant},$$

but no such integration is possible for the equation

$$y_2 \mathrm{d}y_1 + \mathrm{d}y_2 + \mathrm{d}y_3 = 0, \tag{4·8}$$

for which the coefficients do not satisfy (4·6). This does not mean that an adiabatic change governed by an equation such as (4·8) is necessarily impossible—clearly one can always carry out a step-by-step integration of (4·8)—but that what results is not a unique adiabatic surface. It is easy to see in fact that one can connect any two points by means of a line which everywhere satisfies (4·8) (whereas of course if we move so as to satisfy (4·7) we can only connect points which lie on the same sphere). To see this property of (4·8) let us start from the origin of coordinates and move first in the plane $y_2 = \text{constant} = 0$; then $dy_3 = 0$ and the path follows the y_1-axis. Next, starting from any point on the y_1-axis let us move in the plane $y_1 = \text{constant}$; then $dy_2 + dy_3 = 0$, and we move in the plane $y_2 + y_3 = 0$. Thus by a combination of these two progressions we can reach any point in the plane $y_2 + y_3 = 0$. But from any point in this plane we may move so as again to keep y_2 constant, tracing out the straight line $dy_3/dy_1 = -y_2 = \text{constant}$. The family of such lines starting from every point in the plane $y_2 + y_3 = 0$ fills the whole of space, so that we have found a way of connecting all points while still obeying (4·8).

This example will serve to show that we must not, in what follows, take for granted the existence of adiabatic surfaces, except in the special case of simple two-parameter fluids for which the existence of adiabatic lines has been proved. As Carathéodory pointed out, the assumption that adiabatic surfaces exist enables one of the most important consequences of the second law to be deduced purely mathematically. We shall therefore allow ourselves to consider adiabatic changes, but shall not assume that they constitute lines on well-defined adiabatic surfaces until, by means of the second law, we have proved the existence of such surfaces.

For the first development of the consequences of the second law we shall start from Kelvin's formulation, and proceed along the conventional lines, making use of *Carnot cycles*. Consider a system of any degree of complexity, and in particular consider two isotherms of the system which correspond to temperatures θ_1 and θ_2. These isotherms will be lines for simple fluids or 'surfaces' of two or more dimensions for more complex systems. Draw two adiabatic lines which cut the θ_1-surface in A and B, and the θ_2-surface in D and C. The performance of a Carnot cycle then consists of changing the state of the system reversibly from A to B isothermally at θ_1, from B to C adiabatically, from C to D isothermally at θ_2 and finally from D to A adiabatically. Let us suppose that in the isothermal change AB the system takes in an amount of heat Q_1 from a reservoir at θ_1, and along CD an amount Q_2 from another reservoir at θ_2. Then since the system returns to its original state A at the conclusion of the cycle there is no net change in its internal energy, and in consequence it must have performed work equal to $Q_1 + Q_2$.

We may now use Kelvin's law to show that for given values of θ_1 and θ_2 the ratio $-Q_1/Q_2$ is the same for all Carnot cycles, irrespective of the nature of the system concerned. We use a negative sign because Q_1 and Q_2 are necessarily of opposite sign, as follows from Kelvin's law. For if

Q_1 and Q_2 were of the same sign it would be possible to accomplish the cycle in such a sense that both were positive; of the work done, Q_1+Q_2, an amount Q_1 could be put back into the reservoir at θ_1 by some irreversible process, such as friction or Joule heat, and all that would have happened in the cycle would have been the extraction of Q_2 from the reservoir at θ_2 and the performance of work, in contradiction of Kelvin's law.

To show that $-Q_1/Q_2$ takes a constant value for given θ_1 and θ_2, consider two systems performing Carnot cycles. Let the first absorb heats Q_1 and Q_2 in a single cycle, and the second Q_1' and Q_2', and let $n\,|\,Q_1\,|=m\,|\,Q_1'\,|$, where n and m are integers. Then consider the composite system of the two systems taken together, and let a complete cycle of this system consist of n Carnot cycles of the first and m Carnot cycles of the second, the sense of the cycles being chosen that Q_1' and Q_1 are of opposite sign. In this complete cycle the heat absorbed from the reservoir at θ_1, nQ_1+mQ_1', has been made to vanish, while the heat absorbed from the reservoir at θ_2 and converted into work in accordance with the first law, is equal to nQ_2+mQ_2'. By Kelvin's law this cannot be positive,

$$nQ_2+mQ_2'\leqslant 0. \tag{4·9}$$

But since everything is reversible we may carry out the whole cycle backwards, and thus prove that

$$-nQ_2-mQ_2'\leqslant 0, \tag{4·10}$$

from which it follows that

$$nQ_2=-mQ_2'.$$

But by definition

$$nQ_1=-mQ_1'.$$

Hence

$$-Q_1/Q_2=-Q_1'/Q_2',$$

which proves the proposition.

We may therefore infer the existence of a universal function $f(\theta_1,\theta_2)$ having the property that

$$-Q_1/Q_2=f(\theta_1,\theta_2) \quad \text{for a Carnot cycle.} \tag{4·11}$$

Further, we may readily show that $f(\theta_1,\theta_2)$ is decomposible into the quotient $\phi(\theta_1)/\phi(\theta_2)$, where $\phi(\theta)$ is some function of empirical temperature. Since $f(\theta_1,\theta_2)$ is a universal function it is only necessary to prove the decomposition in any one special case in order to establish its generality, and the special case which we shall choose is a simple fluid, for which isotherms and unique adiabatics may be drawn, as in fig. 8.

Here AB, FC, ED are portions of isotherms corresponding to the temperatures θ_1, θ_2 and θ_3 respectively, while AFE and BCD are adiabatics. Let the heat absorbed in going from left to right along the three isotherms be Q_1, Q_2 and Q_3 respectively. Then according to (4·11),

$$Q_1/Q_2 = f(\theta_1, \theta_2),$$

$$Q_2/Q_3 = f(\theta_2, \theta_3)$$

and

$$Q_1/Q_3 = f(\theta_1, \theta_3).$$

Hence

$$f(\theta_1, \theta_3) = f(\theta_1, \theta_2) f(\theta_2, \theta_3). \tag{4·12}$$

Since θ_2 is an independent variable which appears only on the right-hand side of (4·12), the function $f(\theta_1, \theta_2)$ must take the form $\phi(\theta_1)/\phi(\theta_2)$ so that θ_2 may vanish from the equation.

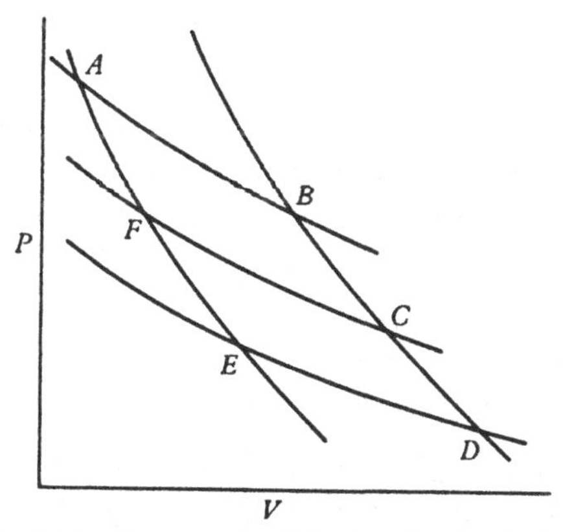

Fig. 8. Isotherms and adiabatics of a simple fluid.

This argument justifies the introduction of the *absolute temperature scale* by the definition of T as proportional to the new function $\phi(\theta)$, or, from (4·11), by writing

$$-Q_1/Q_2 = T_1/T_2 \quad \text{for any Carnot cycle.} \tag{4·13}$$

This equation determines the absolute scale (so called because it does not depend upon the properties of any particular substance, but only on a general property of Carnot cycles) except for an arbitrary constant of proportionality. This constant is normally fixed by the subsidiary requirement that there shall be 100° between the melting-point of ice and the boiling-point of water.† It would in principle be

† It has recently been agreed internationally to abandon the boiling-point as a fixed point and to define the absolute degree in such a way that the ice-point is exactly 273·15° K. This means that the boiling-point is very nearly, but not precisely, 373·15° K. The symbol K (for Kelvin) is used to designate the absolute scale of temperature.

possible to determine the form of the function $\phi(\theta)$ for any given empirical scale by the performance of Carnot cycles under as nearly ideal conditions as possible and the determination of $-Q_1/Q_2$. Since, however, it is possible to show that $T=\theta$ if the perfect gas scale is used to define θ, this tedious process is unnecessary. We shall defer a proof of this result until the analytical development of the second law has proceeded far enough to make it a very straightforward matter.

We have shown that $\Sigma Q/T=0$ in a Carnot cycle, and may now proceed to establish that a similar result holds for any reversible cycle. Consider a system, σ, which executes a cyclical process of any degree of complexity. In the course of traversing one element of this cycle work will be done in general on or by the system, and it will also be necessary to transfer heat to, or abstract heat from, the system. It is convenient to imagine that each element of heat q is transferred to the system from a subsidiary body, σ', at temperature T, which may be caused to execute Carnot cycles between the temperature T and the fixed temperature, T_0, of a heat reservoir. The traversal of the element of the main cycle may then be considered as involving the following procedure:

(1) σ is in its initial state; σ' is at temperature T_0.

(2) σ' is brought adiabatically to T.

(3) σ moves to its final state, absorbing heat q from σ', which proceeds along an isothermal at temperature T.

(4) σ' is returned adiabatically to T_0 and compressed or expanded isothermally until it regains its initial state exactly.

Since in this infinitesimal process σ' loses heat q at temperature T in stage 3, it must gain from the reservoir in stage 4 heat amounting to qT_0/T, by (4·13). In the course of the whole cycle of σ, then, the heat lost by the reservoir is equal to $T_0\oint q/T$, the integral being taken around the cycle of σ. Now the internal energies of both σ and σ' are the same at the end as at the beginning of the cycle, so that this heat, if positive, is equal to the work done by σ and σ' during the course of the cycle. It follows, then, from Kelvin's law that (since T_0 is constant),

$$\oint q/T \leqslant 0. \tag{4·14}$$

If the cycle of σ is reversible we may imagine the whole process carried out backwards, when for every element q in the original cycle we have now $-q$. To avoid violation of Kelvin's law it must now be true that

$$-\oint q/T \leqslant 0.$$

Hence for a reversible cycle

$$\oint q/T = 0. \tag{4·15}$$

It should be noted that T here is strictly the temperature of the body σ' which supplies heat q; but for the cycle to be reversible σ and σ' must have the same temperature when any heat transfer occurs, so that T may be also interpreted as the temperature of the system σ which receives heat q. If, however, the cycle of σ is not reversible then we can only establish the inequality (4·14), which is known as *Clausius's inequality*. And in these circumstances we must be careful to interpret T as the temperature of the body which supplies the heat. This distinction will be of importance when we come to consider cyclical changes in which the system σ may not be in equilibrium at all times, and may not have a single definable temperature.

We shall not at present discuss the consequences of Clausius's inequality, but consider first the important consequences arising from (4·15), which is applicable to reversible processes only. It is an immediate corollary of (4·15) that for all reversible paths between two equilibrium states (A and B) of a system, the integral $\int_A^B q/T$ takes the same value. For if we consider two such paths we may traverse one in the opposite sense from the other and thus construct a reversible cycle. Hence

$$\oint q/T = \int_{A_{\text{path 1}}}^{B} q/T + \int_{B_{\text{path 2}}}^{A} q/T = 0,$$

which proves the proposition. If then we introduce a function S, the *entropy*, by means of the definition

$$S_B - S_A = \int_A^B q/T \quad \text{for a reversible change from } A \text{ to } B, \tag{4·16}$$

or, differentially,

$$\mathrm{d}S = q/T \qquad \text{for an infinitesimal reversible change,} \tag{4·17}$$

it follows that the difference $S_B - S_A$ is independent of the path chosen to connect A and B, and that S is therefore a function of state, completely determinable once its value has been arbitrarily fixed for one particular state of the system. This result enables us to infer immediately that adiabatic surfaces exist for all systems, however complex. For in an adiabatic (reversible) change $Q=0$ and hence S is constant; the family of surfaces $S=$ constant constitutes the adiabatic surfaces of the system.

In deriving the existence of adiabatic surfaces and of the entropy function, we have used the second law liberally in order to avoid mathematical argument. Let us now see how we can arrive at the same result by a more economical use of physical principles and a more lavish use of mathematics. We shall still employ Kelvin's formulation of the second law, but shall first establish the existence of adiabatic surfaces, without employing Carnot cycles, and thence deduce the existence of entropy and of an absolute scale of temperature. The argument is more readily visualized if we consider a system defined by three parameters, for which the isothermals are two-dimensional surfaces. Let us construct an adiabatic surface by the following procedure. Draw any adiabatic line, i.e. one which at all points satisfies the equation (which is the same as (4·5))

$$Y_1 \mathrm{d}y_1 + Y_2 \mathrm{d}y_2 + Y_3 \mathrm{d}y_2 = 0. \tag{4·18}$$

Starting from each point on this line construct an adiabatic line which lies on the isothermal surface through that point. In this way we have constructed an adiabatic surface, and we must now show, by use of Kelvin's law, that our procedure has led to a unique surface satisfying (4·18) at all points. To do this we first consider two points, P and P', lying in the surface and infinitesimally close to one another. It is possible to join these by means of a curve formed from parts of the adiabatic lines whereby the surface was constructed, so that along this curve $\int_P^{P'} \mathrm{d}Q = 0$. Hence the infinitesimal element PP' must satisfy (4·18). For if it did not it would satisfy an equation of the form

$$Y_1 \mathrm{d}y_1 + Y_2 \mathrm{d}y_2 + Y_3 \mathrm{d}y_3 = \mathrm{d}Q$$

with a non-zero value of $\mathrm{d}Q$; this would imply that a reversible cycle could be constructed in which heat $\mathrm{d}Q$ was absorbed from a reservoir at the temperature corresponding to the point P, and an equal amount of work done, in violation of Kelvin's law. Thus all line elements in the surface are solutions of (4·18), so that all curves lying wholly in the surface are adiabatics. Moreover, since the tangent plane to the surface at P has now been shown to be a solution of (4·18), it follows that an infinitesimal line element PP'', joining P to any neighbouring point P'' not lying in the surface, is not a solution of (4·18) and hence PP'' is not an adiabatic line; by the same argument as before we conclude that there can be no adiabatic line, however roundabout, connecting P and P''. The surface we have constructed is thus entirely surrounded by adiabatically inaccessible points, and so is a unique adiabatic surface. The argument may readily be extended to systems of any number of parameters.

Knowing that adiabatic surfaces exist, we may now prove without further reference to the second law, that in a reversible change $\mathrm{d}Q$ may be written as $T\mathrm{d}S$, in which T is a function of empirical temperature only, and S is a function of state. To do so we imagine each separate adiabatic surface of a system having n independent parameters to be arbitrarily labelled with a number ϕ, and thus introduce a new function of state $\phi(y_i)$ which has the property that $\phi = \text{constant}$ for an adiabatic change, or

$$\sum_i \frac{\partial \phi}{\partial y_i} \mathrm{d}y_i = 0. \tag{4·19}$$

In addition, for an adiabatic change, (4·3) holds:

$$\sum_i Y_i \,\mathrm{d}y_i = 0.$$

Therefore, if $\lambda(y_i)$ is any function of state, it follows from (4·3) and (4·19) that on an adiabatic surface

$$\sum_i \left(Y_i - \lambda \frac{\partial \phi}{\partial y_i} \right) \mathrm{d}y_i = 0. \tag{4·20}$$

Now of all the n coordinates y_i, only $(n-1)$ are independent, since the $\mathrm{d}y_i$ must satisfy (4·19). Let us regard y_1 as the dependent variable, and choose λ such that

$$Y_1 - \lambda \frac{\partial \phi}{\partial y_1} = 0.$$

The first term in (4·20) now vanishes and we are left with an equation involving only independent variables. If then (4·20) is to be generally true, each coefficient of $\mathrm{d}y_i$ must vanish separately, and

$$Y_i = \lambda \frac{\partial \phi}{\partial y_i} \quad \text{for all } i. \tag{4·21}$$

Hence in a reversible change which is not adiabatic,

$$\mathrm{d}Q = \sum_i Y_i \mathrm{d}y_i = \sum_i \lambda \frac{\partial \phi}{\partial y_i} \mathrm{d}y_i = \lambda \,\mathrm{d}\phi. \tag{4·22}$$

It remains to demonstrate that λ is a function of θ and ϕ only. For this purpose consider a composite system consisting of two separate systems in thermal equilibrium, at a common empirical temperature θ, the first system having n coordinates $y_1 \ldots y_n$, and the second m coordinates $z_1 \ldots z_m$. For each system we may introduce the new functions, $\phi(y_i)$ and $\lambda(y_i)$ for the first, $\phi'(z_i)$ and $\lambda'(z_i)$ for the second, and we have that in a reversible change

$$\mathrm{d}Q = \lambda \,\mathrm{d}\phi \quad \text{for the first system,}$$

$$\mathrm{d}Q' = \lambda' \,\mathrm{d}\phi' \quad \text{for the second system.}$$

For the combined system the same arguments may be used to demonstrate the existence of functions $\Phi(y_i, z_i)$ and $\Lambda(y_i, z_i)$, such that for any reversible change,

$$\Lambda\,\mathrm{d}\Phi = \mathrm{d}Q + \mathrm{d}Q' = \lambda\,\mathrm{d}\phi + \lambda'\mathrm{d}\phi',$$

or

$$\mathrm{d}\Phi = \frac{\lambda}{\Lambda}\mathrm{d}\phi + \frac{\lambda'}{\Lambda}\mathrm{d}\phi'. \tag{4·23}$$

Now since θ, ϕ and ϕ' are all functions of state, we may carry out a change of variables and describe the first system by θ, ϕ and $n-2$ of the original y-coordinates, and the second system by θ, ϕ' and $m-2$ of the original z-coordinates. Then although Φ is in principle a function of all these variables, nevertheless (4·23) shows that there are only two terms in the total differential of Φ; thus Φ is in fact a function of only ϕ and ϕ' and is independent of all the other coordinates, in particular of θ. Hence $\partial\Phi/\partial\phi$, i.e. λ/Λ, and $\partial\Phi/\partial\phi'$, i.e. λ'/Λ, are independent of all coordinates except ϕ and ϕ'. Therefore since λ is independent of the z-coordinates, Λ also must be independent of the z-coordinates; and since λ' is independent of the y-coordinates, Λ also must be independent of the y-coordinates. Hence

$$\Lambda = \Lambda(\phi, \phi', \theta), \quad \lambda = \lambda(\phi, \theta), \quad \lambda' = \lambda'(\phi', \theta).$$

But, further, since λ/Λ and λ'/Λ are functions only of ϕ and ϕ',

$$\frac{\partial}{\partial\theta}\left(\frac{\lambda}{\Lambda}\right) = \frac{\partial}{\partial\theta}\left(\frac{\lambda'}{\Lambda}\right) = 0,$$

i.e.

$$\frac{\partial}{\partial\theta}(\log\lambda) = \frac{\partial}{\partial\theta}(\log\lambda') = \frac{\partial}{\partial\theta}(\log\Lambda). \tag{4·24}$$

In (4·24) we have the first term a function of ϕ and θ only, the second a function of ϕ' and θ only, ϕ and ϕ' being independent variables. Hence in reality each term in (4·24) is a function only of θ. We have therefore proved that for any system

$$\frac{\partial}{\partial\theta}(\log\lambda) = g(\theta), \tag{4·25}$$

in which $g(\theta)$ is a universal function of θ, or, integrating (4·25),

$$\lambda = F(\phi)\exp\left[\int g(\theta)\,\mathrm{d}\theta\right], \tag{4·26}$$

the form of the function $F(\phi)$ being determined by the way in which the adiabatic surfaces were labelled, i.e. by the form chosen for the function ϕ. If now we define

$$T(\theta) \equiv C\exp\left[\int g(\theta)\,\mathrm{d}\theta\right], \tag{4·27}$$

$$S(\phi) \equiv \frac{1}{C}\int F(\phi)\,\mathrm{d}\phi, \tag{4·28}$$

then (4·22) takes the form $dQ = T\,dS$, in which T is a universal function of θ, and S is constant in an adiabatic change. We have thus arrived at the result expressed in (4·17) by a quite different route.

Finally, let us see in outline how Carathéodory's law may be used instead of Kelvin's to establish the existence of adiabatic surfaces. According to this law there are in the neighbourhood of any state other states which are inaccessible by an adiathermal, and *a fortiori* by an adiabatic, process. By the word *neighbourhood* we need not imply that the inaccessible states are infinitesimally close, although that is the conclusion we shall reach eventually. All we mean is that the inaccessible states are always close at hand, and do not consist of a few singular points or surfaces which are inaccessible from everywhere else. If then the nearest adiabatically inaccessible point Q from a given point P is a finite distance from it, we may assume that this holds in general wherever P may be situated. But this is clearly impossible. For if we join P and Q by a line L, there will be on L a point P' which, being nearer than Q, must be accessible from P but inaccessible from Q; since P' may be made as close as we choose to Q it follows that the nearest inaccessible point to P is at a finite distance from P, while the nearest inaccessible point to Q is infinitesimally close. Hence we conclude that Carathéodory's law can be obeyed only if there are adiabatically inaccessible points infinitesimally close to any given point. We may now proceed as in the earlier analysis and, starting from any point P, draw an adiabatic line through P, an adiabatic surface through this line, and so on, according to the number of parameters involved. If we apply the same procedure to two points, Q and Q', infinitesimally close to P on opposite sides of the surface, and inaccessible from P, we may cordon off the surface through P by two inaccessible surfaces as close as we please, and thus demonstrate the uniqueness of the adiabatic surface constructed. From this point onward the rest of the argument proceeds as before.†

We have thus by a variety of means demonstrated that for a reversible change dQ may always be written as $T\,dS$; further, in an irreversible cycle $\oint q/T < 0$, where T is the temperature of the body supplying the element of heat q. We have only proved Clausius's inequality by one means; to construct a proof without invoking cyclical processes may be left as a not very easy exercise for the reader. We shall not at this stage discuss Clausius's inequality further,

† This argument does not do justice to the precise mathematical reasoning of Carathéodory's development, but is intended to show in a simple way how Carathéodory's principle leads to a proof of the existence of adiabatic surfaces. For more careful discussions consult A. H. Wilson, *Thermodynamics and Statistical Mechanics* (Cambridge, 1956), R. Eisenschitz, *Sci. Progr.* **170**, 246 (1955).

but first confine our attention to the equality applicable to reversible changes. Since in a reversible change, according to the first law,

$$\mathrm{d}U = \mathrm{d}Q + \sum_i X_i \,\mathrm{d}x_i,$$

we have that

$$\mathrm{d}U = T\,\mathrm{d}S + \sum_i X_i \,\mathrm{d}x_i. \tag{4·29}$$

This result, derived by a consideration of reversible changes, states a relationship between functions which are all functions of state. We may therefore make a most important extension of the range of validity of (4·29), and declare that it is applicable to *any* differential change, reversible or irreversible. For example, for any change in a fluid,

$$\mathrm{d}U = q + w \tag{4·30}$$

and

$$\mathrm{d}U = T\,\mathrm{d}S - P\,\mathrm{d}V. \tag{4·31}$$

It is only for a reversible change, however, that $q = T\,\mathrm{d}S$ and $w = -P\,\mathrm{d}V$; neither of these equalities hold for an irreversible change—$q \neq T\,\mathrm{d}S$ and $w \neq -P\,\mathrm{d}V$—but (4·31) is still valid; if $q = T\,\mathrm{d}S - \epsilon$ then $w = -P\,\mathrm{d}V + \epsilon$. This point is illustrated by the experiment pictured in fig. 3, the expansion of gas into a vacuum under isolated conditions. Here $P\,\mathrm{d}V$ has a definite, non-zero value, but $w = 0$; similarly $q = 0$ but $T\,\mathrm{d}S$ has a non-zero value, which could be determined by compressing the gas reversibly to its initial state and determining how much heat must be extracted during the compression. We should find that it was just equal to $P\,\mathrm{d}V$, and that in consequence the entropy increase during the irreversible expansion was $P\,\mathrm{d}V/T$.

We have now solved the problem which we set ourselves, of finding expressions for q and w in terms of functions of state, so that we should have only exact differentials to handle. Equation (4·29) is the solution, valid for all changes, and it is from this equation that the analytical application of thermodynamics to physical problems stems. The following chapters will be concerned with such applications.

CHAPTER 5

A MISCELLANY OF USEFUL IDEAS

Dimensions and related topics

In the last chapter we derived the important result that for any differential change

$$\mathrm{d}U = T\mathrm{d}S + \sum_i X_i \mathrm{d}x_i. \tag{5·1}$$

In particular, for a fluid

$$\mathrm{d}U = T\mathrm{d}S - P\mathrm{d}V, \tag{5·2}$$

and corresponding results hold for other two-parameter systems in which P and V are replaced by the appropriate X and x; for instance, a solid magnetizable body may often be regarded as unaffected by pressure changes and under this assumption it obeys the equation

$$\mathrm{d}U' = T\mathrm{d}S + \mathscr{H}\,.\,\mathrm{d}\mathscr{M}. \tag{5·3}$$

We shall consider the consequences of (5·2); they can obviously be applied, *mutatis mutandis*, to (5·3) or similar modifications.

If any thermodynamic variable is a function of two parameters, it is possible to consider any two functions of state as independent variables, and we have available for a fluid the variables, U, P, V, T, S, or combinations of these such as H the *enthalpy*, defined as $U+PV$; F the *Helmholtz free energy*, defined as $U-TS$; G the *Gibbs function*, defined as $U-TS+PV$; or any other suitable combinations. It is desirable of course that the combination chosen should be dimensionally homogeneous, and clearly this condition is satisfied by H, F and G, which all have the dimensions of energy. In this connexion it is worth remarking on three points. First, there is nothing in the way temperature and entropy are introduced which enables us to determine their dimensions separately; all we can say is that the product TS has the same dimensions as U. We must expect therefore that S will never appear in any equation without T to render it dimensionally meaningful, unless the equation is in itself homogeneous in S so that its dimensions are unimportant. The second point concerns the variation of the thermodynamic parameters with the size of the body considered. Unless surface phenomena are concerned the behaviour under equilibrium conditions of similar bodies in given circumstances is not dependent on their size, and the work required to cause a given change of temperature, say, to a body of given constitution under conditions of thermal isolation is simply proportional to the size of the body. It follows that U is proportional to the size of the body, and we shall therefore expect TS and PV (or any $X_i x_i$) to be proportional to the size, otherwise (5·1) would predict size-dependent behaviour. It is clear that this proportionality is usually achieved by having only one member of each pair of variables size-dependent. There are variables such as T, P or $\mathscr{H}$ (*intensive* variables) which determine the conditions to which the body is subjected, and which are size-independent, and there are the corresponding variables S, V and $\mathscr{M}$ (*extensive* variables) which under given conditions are proportional to the size of the body. In these examples the intensive variables have happened to be the coefficients of the differentials in (5·1), but this is not a general rule. For example, if we consider a wire in tension we cannot ascribe the terms intensive or extensive uniquely, since the behaviour of the wire depends not simply on its volume but on its shape as well. If we are concerned only with wires of a given cross-sectional area, the work term if written $f\mathrm{d}l$ has f as an intensive and l as an extensive variable, but if it is written $fl\mathrm{d}\epsilon$ ($\mathrm{d}\epsilon \equiv \mathrm{d}l/l$), then according to our definition fl is extensive and ϵ intensive. We need not concern ourselves with such hair-splitting, however; in any practical application no ambiguity arises.

Finally, we may note that if we have a composite system, consisting of several subsystems, each in equilibrium within itself, though, if thermally isolated, not necessarily in equilibrium with the others, the values of the extensive variables V, U and S for the whole system may be taken without inconsistency to be simply the sum of the contributions from all the subsystems. For V this is obvious; proof for the other variables, based on the definitions of ΔU and ΔS, is left to the reader. The proviso that each subsystem shall be in equilibrium is of course essential if the entropy is to have any meaning. This additivity rule applies also to the free energy F if all the subsystems have the same temperature, since

$$F = U - TS = \Sigma U_i - T\Sigma S_i = \Sigma(U_i - TS_i) = \Sigma F_i,$$

the summations being taken over all subsystems. But if the subsystems are not at the same temperature the free energy of the system as a whole is not a meaningful concept (since the temperature of the system as a whole is undefined); it may, however, be arbitrarily defined as ΣF_i. Similarly, the additivity rule applies to the Gibbs function G only if T and P are constant throughout the system, though again, if required, G may be defined as ΣG_i even in non-uniform systems.

Maxwell's thermodynamic relations

Returning now to (5·2), we may regard U and all other parameters as functions of two independent variables, say y and z, and in general

$$\left.\begin{aligned}\left(\frac{\partial U}{\partial y}\right)_z&=T\left(\frac{\partial S}{\partial y}\right)_z-P\left(\frac{\partial V}{\partial y}\right)_z\\ \text{and}\qquad \left(\frac{\partial U}{\partial z}\right)_y&=T\left(\frac{\partial S}{\partial z}\right)_y-P\left(\frac{\partial V}{\partial z}\right)_y.\end{aligned}\right\}\tag{5·4}$$

In particular, we may choose either or both of S and V as our independent variables. If S and z are chosen,

$$\left(\frac{\partial U}{\partial S}\right)_z=T-P\left(\frac{\partial V}{\partial S}\right)_z;\tag{5·5}$$

if S and V are chosen,

$$\left(\frac{\partial U}{\partial S}\right)_V=T\tag{5·6}$$

and

$$\left(\frac{\partial U}{\partial V}\right)_S=-P.\tag{5·7}$$

From (5·6) and (5·7) an important result is obtained by a second differentiation. From (5·6)

$$\frac{\partial^2 U}{\partial V\partial S}=\left(\frac{\partial T}{\partial V}\right)_S,$$

and from (5·7)

$$\frac{\partial^2 U}{\partial S\partial V}=-\left(\frac{\partial P}{\partial S}\right)_V.$$

Since in the double differentiation the order of differentiation is immaterial, we have that

$$\left(\frac{\partial T}{\partial V}\right)_S=-\left(\frac{\partial P}{\partial S}\right)_V.\tag{M.1}$$

This equation is the first of Maxwell's four thermodynamic relations. In all that follows these relations will be referred to as M. 1, 2, 3 and 4. The other three may be obtained by use of the functions H, F and G. For example, since $H=U+PV$, in general

$$\begin{aligned}\mathrm{d}H&=\mathrm{d}U+P\,\mathrm{d}V+V\,\mathrm{d}P\\ &=T\,\mathrm{d}S+V\,\mathrm{d}P\quad\text{from (5·2).}\end{aligned}$$

Hence by double differentiation, taking S and P as independent variables, we have that

$$\left(\frac{\partial T}{\partial P}\right)_S=\left(\frac{\partial V}{\partial S}\right)_P.\tag{M.2}$$

The reader may easily verify in a similar manner the last two:

$$\left(\frac{\partial V}{\partial T}\right)_P = -\left(\frac{\partial S}{\partial P}\right)_T \tag{M.3}$$

and

$$\left(\frac{\partial P}{\partial T}\right)_V = \left(\frac{\partial S}{\partial V}\right)_T. \tag{M.4}$$

These four equations should not be regarded as independent deductions from (5·2); given one, the others may be deduced by manipulation alone, by means of the mathematical identities for functions of two variables:

$$\left(\frac{\partial z}{\partial x}\right)_y \left(\frac{\partial x}{\partial y}\right)_z \left(\frac{\partial y}{\partial z}\right)_x = -1 \tag{5·8}$$

and

$$\left(\frac{\partial w}{\partial x}\right)_y \Big/ \left(\frac{\partial z}{\partial x}\right)_y = \left(\frac{\partial w}{\partial z}\right)_y. \tag{5·9}$$

To show this, let us start from M.1, which may be rewritten by use of (5·8),

$$\left(\frac{\partial T}{\partial V}\right)_S = \left(\frac{\partial P}{\partial V}\right)_S \left(\frac{\partial V}{\partial S}\right)_P,$$

and by use of (5·9),

$$\left(\frac{\partial T}{\partial P}\right)_S = \left(\frac{\partial V}{\partial S}\right)_P,$$

which is the same as M.2. Repetition of the procedure yields M.3 and M.4.†

Identity of the absolute and perfect gas scales of temperature

The chief value of Maxwell's relations is that they allow a rearrangement of differential coefficients so that the relationship between different observable phenomena is exhibited. This is particularly true

† In view of the wide application of Maxwell's relations it is worth while committing them to memory, even though their derivation is so simple. The following remarks may help in remembering them. First we may note that they are dimensionally homogeneous, in that cross-multiplication yields each time the pairs TS and PV (the operator ∂ is of course dimensionless). Secondly, the equations may always be written so as to exhibit the independent variables in the denominator. Maxwell's relations represent the four possible equations which satisfy these requirements, so that one can write them down immediately, complete except for the signs, of which two are positive and two negative. The signs can easily be found by inspection if we consider the meaning of the equations when they are applied to a perfect gas. To take M.1 as an example, the left-hand differential coefficient describes the increase in temperature when the gas is expanded adiabatically, and is therefore negative, since a gas cools on expansion; the right-hand differential coefficient describes the increase in pressure when a gas is heated at constant volume, and is therefore positive. The necessity for the negative sign is now clear. The reader should satisfy himself that he understands the meaning of the other three relations in the same way.

of M.3 and M.4, which contain on the right-hand side quantities which are difficult to measure in practice, as they involve calorimetry, while the quantities on the left-hand side may be computed from the equation of state, which may usually be found experimentally without excessive difficulty. This point should become clear as we proceed, in later chapters, to the applications of thermodynamics. A simple example of the use of Maxwell's relations is, however, conveniently introduced here, to justify an assertion made earlier that the perfect gas scale and the thermodynamic scale of temperature are identical. We must first define what we mean by a perfect gas without reference to a scale of temperature, and for this purpose we make use of the following experimental facts:

(1) As the pressure of a real gas tends to zero, the product PV at constant temperature tends to a finite limit (Boyle's law).

(2) As the pressure of a real gas tends to zero, the internal energy at constant temperature tends to a finite limit (Joule's law).

The experimental evidence for the second statement is, first, Joule's observation that a gas expanding into a vacuum under isolated conditions experiences no change of temperature, and, secondly, more recent experiments† in which more precise observation has revealed a dependence of U on P which becomes negligible as the pressure is lowered. We take the laws of Boyle and Joule to define a perfect gas, in confidence that a good approximation to a perfect gas can be found in practice, and that the behaviour of perfect gases may in principle be found experimentally by extrapolating the behaviour of real gases to zero pressure (see p. 89). The two laws are sufficient to enable us to deduce the equation of state of a perfect gas thermodynamically.

For any gas,

$$dU = T\,dS - P\,dV,$$

so that

$$\left(\frac{\partial U}{\partial V}\right)_T = T\left(\frac{\partial S}{\partial V}\right)_T - P$$

$$= T\left(\frac{\partial P}{\partial T}\right)_V - P \quad \text{from M.4.} \qquad (5{\cdot}10)$$

In particular, for a perfect gas $(\partial U/\partial V)_T = 0$ from Joule's law, so that

$$\left(\frac{\partial P}{\partial T}\right)_V = \frac{P}{T},$$

or $P = TF(V)$, where $F(V)$ is an as yet unknown function of V. But according to Boyle's law, at constant temperature $P \propto 1/V$. Therefore $F(V) \propto 1/V$, and $PV = kT$, where k is a constant. Now the perfect gas scale of temperature is defined by the equation $PV = R\theta$, so that θ/T is

† F. D. Rossini and M. Frandsen, *J. Res. Nat. Bur. Stand.* **9**, 733 (1932).

a constant. By choosing the same fixed points the constant can be made to equal unity, and the scales of temperature coincide. This argument does not of course solve the practical problem of calibrating a thermometer according to the absolute scale; it indicates a method whereby it can be done. The experimental problems of constructing a gas thermometer and extrapolating its behaviour to zero pressure are very small if only moderate accuracy is desired, but if uncertainties of $\frac{1}{100}°$ or less are sought very elaborate precautions are needed. We shall not enter upon this problem at all here, but refer the interested reader to other treatises for more detailed accounts.† Some of the thermodynamical problems involved in establishing the absolute scale will be discussed more fully in the next chapter.

Absolute zero, negative temperatures and the third law of thermodynamics

Measurements with gas thermometers, and by other means, have established that the melting-point of ice on the absolute Centigrade scale of temperature is 273·15° K. By the use of gas liquefiers lower temperatures than this are readily attained, the normal boiling-point of liquid helium, for example, being 4·2° K. A bath of liquid helium, contained in a Dewar vessel to minimize the leakage of heat from outside, may be cooled further by pumping away the vapour and causing the liquid to boil at a reduced pressure. In this way temperatures as low as 0·8° K. are attainable, but it is not feasible to go much lower than this since the vapour pressure of helium falls very rapidly towards zero below 0·8° K., and no pump will maintain a low enough pressure against the large volume of gas which evaporates from the liquid under those conditions. To reach still lower temperatures the technique of *adiabatic demagnetization* was devised (p. 67), and the lowest temperature yet reached by this means is about 0·001° K. These experiments in themselves suggest that, however low the temperature may be brought, there may be some limitation to all cooling methods which prevents the *absolute zero*, 0° K., from ever being reached.

It was once a rather widely held belief that the second law provided arguments against the possibility of attaining the absolute zero, and we shall now indicate the lines of thought involved, although as we shall see later they cannot be regarded as convincing. The most efficient way of lowering temperature is to employ an adiabatic process, such as expanding an isolated mass of gas reversibly, lowering the pressure over an isolated bath of liquid, or reducing the magnetic field applied to an isolated block of paramagnetic salt. All these methods depend

† *Temperature, Its Measurement and Control in Science and Industry* (Reinhold, 1941). F. Henning, *Temperaturmessung* (Barth, 1951).

for their operation on the use of a substance or system in which the entropy may be varied at constant temperature by the change of some convenient parameter β (e.g. V, P and $\mathscr{H}$ in the three examples above). For if $(\partial S/\partial\beta)_T$ does not vanish, then by (5·8) neither does $(\partial S/\partial T)_\beta(\partial T/\partial\beta)_S$; therefore so long as $(\partial S/\partial T)_\beta$ is finite,† an adiabatic change of β will alter the temperature. For a given change of β the greatest cooling is achieved by making the change reversibly, as the reader should be able to prove from the second law.

Now suppose that by such an adiabatic process the temperature of the system may be reduced to zero. It might appear that this provides the possibility of operating a Carnot engine between zero and a non-zero temperature, and that this would enable the second law to be violated, since such an engine need discharge no heat at the lower operating temperature and can still do useful work. But we must be careful in jumping to this conclusion, since a Carnot cycle involves the performance of an isothermal change of entropy at the lower operating temperature, and if this is zero the isothermal change is also adiabatic, in the sense that no heat is needed to alter the entropy at 0° K. This is, however, to some extent only an ambiguity of terms, for the essential distinction to be made is between a reversible adiathermal change and an isentropic change, which are of course identical except at 0° K. If a reversible change in the value of β at 0° K. alters the entropy, then the two adiabatic portions of the Carnot cycle are performed at different values of the entropy, and this enables heat to be taken in at the upper operating temperature, and the second law to be violated. On the other hand, if the entropy is not dependent upon β at 0° K. the two adiabatic portions necessarily coincide, the cycle is inoperative, and the validity of the second law is preserved.

It has been argued in this way that the second law requires that any reversible change taking place at the absolute zero shall involve no change in entropy; and this requirement is readily seen to preclude the attainment of the absolute zero. For if we construct adiabatic surfaces in the coordinate system of T and any relevant parameters (X, x), the best hope of reaching 0° K. is to move on one of these

† If real physical systems were governed by the laws of classical mechanics the problem of trying to reach 0° K. would never arise. For Boltzmann's law of equipartition of energy would ensure that the specific heat remained at a non-vanishing value at all temperatures. Thus as T tended to zero $(\partial S/\partial T)_\beta$ would become infinite as T^{-1}. In other words, the entropy of all substances would tend to $-\infty$ as T tended to zero, and no isentropic process could reduce T to zero. In fact, however, it is a consequence of the quantal behaviour of matter that all specific heats tend to zero at 0° K. at least as fast as T (for metals $C_V \propto T$, for non-metals and liquid helium $C_V \propto T^3$ at the lowest temperatures). Thus the entropy tends to a finite limit and the question of the attainability of absolute zero is one which must be considered.

surfaces from some starting temperature T_0 and finish eventually at 0° K. But if the above argument is valid, a single adiabatic surface covers the whole variation of the parameters (X, x) at 0° K., and therefore the only chance of reaching 0° K. is for this surface (on which we may for convenience put $S=0$) to have a branch into the region of higher temperatures, as is indicated for one parameter β in fig. 9. It will now be seen that this automatically involves $(\partial S/\partial T)_\beta$ and therefore C_β becoming negative in some region of the diagram. However, C_β being a principal specific heat is always positive, as pointed out on p. 18, and therefore the zero entropy contour cannot branch as in fig. 9 and the absolute zero is unattainable.

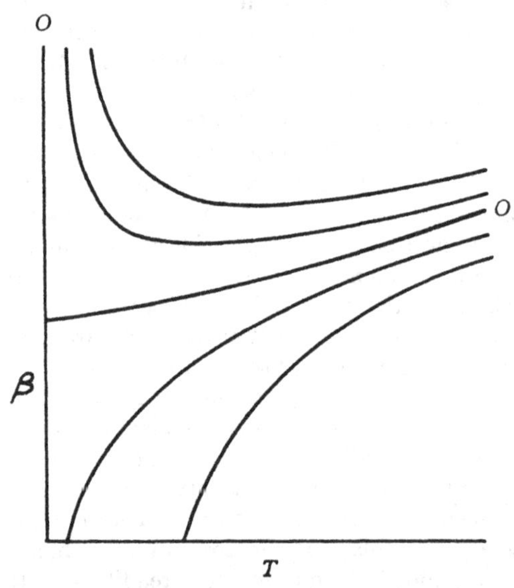

Fig. 9. Hypothetical branched adiabatic.

Three criticisms may be directed against the foregoing argument. The first is that we have considered an idealized experiment, in that we have postulated perfect reversibility in the Carnot cycle, and perfect isolation in the adiabatic parts. Although this may be legitimate at non-zero temperatures, since, for example, there is a meaning to be attached to the limit, as δQ goes to zero, of the entropy change $\delta Q/T$ resulting from a leakage of heat, this limit is meaningless when $T=0$. Any heat leak, however small, from the surroundings to the system at 0° K., raises its temperature above 0° K. by an amount which is never negligible. Secondly, it may be argued that it is taking the second law too far beyond the range of its experimental foundations to apply it in the vicinity of 0° K.; it might be that the Carnot engine considered constitutes the one exception to its general validity. But the third criticism is the most cogent. We have assumed that if the system is such that $(\partial S/\partial \beta)_T$ does not vanish at 0° K., it is possible to operate a Carnot cycle without a reservoir at the lower operating temperature, 0° K. Certainly this is true in the sense already discussed, that an isothermal change at zero temperature involves no transfer of heat. But this very circumstance makes the cycle inoperative; for there is no practical way of compelling the isolated system to perform the isothermal change rather than an adiabatic (isentropic) change. The ambiguity of adiabatic and isothermal changes at 0° K., out of which we attempted to talk our way on p. 49, is thus a crucial matter which invalidates any attempt along these lines to demonstrate that the unattainability of the absolute zero is a consequence of the second law.

In view of this failure we must, if we are to incorporate the idea in the development of thermodynamics, introduce it as a new postulate, the third law of thermodynamics:

By no finite series of processes is the absolute zero attainable.

By reversing the argument given above it is now readily seen that no isentropic surface connects any point at 0° K. with any other point at a higher temperature, and therefore, since S remains finite down to 0° K., all points at 0° K. lie on a single isolated isentropic surface. This enables us to give an alternative statement to the third law:

As the temperature tends to zero, the magnitude of the entropy change in any reversible process tends to zero.

We shall meet, in the following chapters, a few examples which demonstrate the application of the third law, but it has not the same importance in physics as it has in chemistry, where it plays a most valuable role in enabling the equilibrium constants of chemical reactions to be calculated from the thermal properties of the reactants. The success which attends its application in this field leaves no room for doubt of its correctness. In view, however, of the limited use we shall make of the law, and the need for a full development of chemical thermodynamics in order to appreciate its application, we shall not discuss it any further.

From the unattainability of the absolute zero and the experimental fact (which is a consequence of the validity of classical mechanics in describing the atomic behaviour of matter at high temperatures) that the specific heat of a complete system does not tend to zero as T tends to infinity, it follows that the temperatures of all bodies have the same sign, which is positive by definition. Recently an experiment was reported in which part of a system (the atomic nuclei of a solid) was brought by magnetic means into what may be described with some justification as a state of negative temperature.† A full discussion of this experiment lies outside the realm of classical thermodynamics, since it involves a microscopic viewpoint to see the assembly of atomic nuclei as forming a subsystem which can be considered isolated from the rest of the solid lattice. The specific heat of such a subsystem tends to zero at high temperatures sufficiently rapidly for only a finite amount of energy to be needed to raise the temperature to infinity. The thermal conductance between the nuclear subsystem and the rest of the lattice is so low that this may be achieved while the lattice is maintained at ordinary temperatures. It is even possible to add still more energy to the subsystem, and this is equivalent to forcing its temperature into the negative region. A microscopic analysis, by statistical means, shows that $+\infty$ and $-\infty$ are indistinguishable as temperatures. This is because from the point of view of statistical thermodynamics it is $1/T$, rather than T, which is the physically significant parameter, and there is no energetic barrier in these particular experiments preventing passage through the origin of $1/T$. Thus negative temperatures are hotter than any positive temperature, and the hottest possible temperature is -0, while the coldest is $+0$. This experiment shows very clearly the essentially singular nature of the absolute zero, which is quite impassable by any means. It is legitimate to inquire whether the second law may be violated by working a Carnot engine between two temperature baths of which one is negative and the other positive. It may be shown that no violation is possible, since no isentropic surfaces connect positive and negative temperatures, and therefore no reversible cycle may be constructed. In the experiment the passage from positive to negative temperature was effected by an ingenious trick which does not correspond to any ordinary reversible process.

This experiment has been mentioned solely in order to point out that there are exceptions to the rule that temperatures are positive. But these exceptions never occur with complete systems in equilibrium, only for very special isolable subsystems, and for normal purposes we take T to be always a positive quantity.

† The reader cannot be expected to understand the nature of this experiment without a rather detailed knowledge of the field of *spin resonance*. The experiment is described and discussed by N. F. Ramsey, *Phys. Rev.* **103**, 20 (1956).

An elementary graphical method of solving thermodynamic problems

We shall now examine briefly a graphical method which is sometimes used to solve elementary thermodynamic problems. The substance considered is imagined taken around a Carnot cycle between two neighbouring temperatures T and $T-\delta T$. If the area of the cycle in the indicator diagram is δA, and the heat absorbed at T is Q, then the heat given out at $T-\delta T$ is $Q-\delta A$, since δA is the work done in the cycle. Hence, by (4·13),

$$\frac{Q-\delta A}{Q}=\frac{T-\delta T}{T},$$

or

$$\delta A/Q=\delta T/T. \tag{5·11}$$

Evaluation of the quantities occurring in (5·11) leads to the desired result. We may illustrate the method by deriving Clapeyron's equation for the variation of vapour pressure of a liquid with temperature. In fig. 10 are drawn two neighbouring isotherms of a liquid-vapour

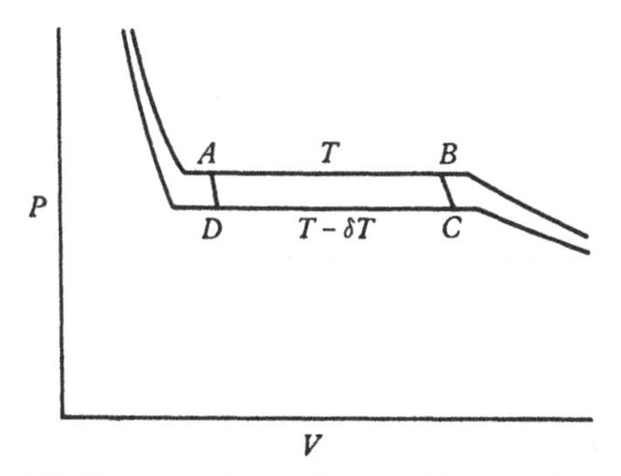

Fig. 10. Carnot cycle for deriving Clapeyron's equation.

system, the horizontal portions corresponding to the two phases coexisting in equilibrium. The Carnot cycle consists of two long isotherms AB and CD and two short adiabatics BC and DA. We suppose the variation from A to B to correspond to the evaporation at constant pressure P of unit mass of liquid, necessitating the absorption of heat equal to the latent heat per unit mass l. To find the area of the cycle we note that the length AB is $v_v - v_l$, the difference in volume between unit mass of vapour and liquid, while the vertical width of the cycle is $(\mathrm{d}P/\mathrm{d}T)\,\delta T$, P being the equilibrium vapour pressure. Thus

$\delta A = (v_v - v_l)(\mathrm{d}P/\mathrm{d}T)\,\delta T$, any difference in slope of AD and BC becoming negligible as $\delta T \to 0$. Hence from (5·11),

$$\frac{1}{l}(v_v - v_l)\left(\frac{\mathrm{d}P}{\mathrm{d}T}\right)\delta T = \delta T/T,$$

or

$$\frac{\mathrm{d}P}{\mathrm{d}T} = \frac{l}{T(v_v - v_l)}, \tag{5·12}$$

which is Clapeyron's equation, expressing the variation of vapour pressure with temperature in terms of other measurable quantities.

Let us now see what this method amounts to, by applying it to an infinitesimal Carnot cycle of a simple fluid. In fig. 11, AB and CD are

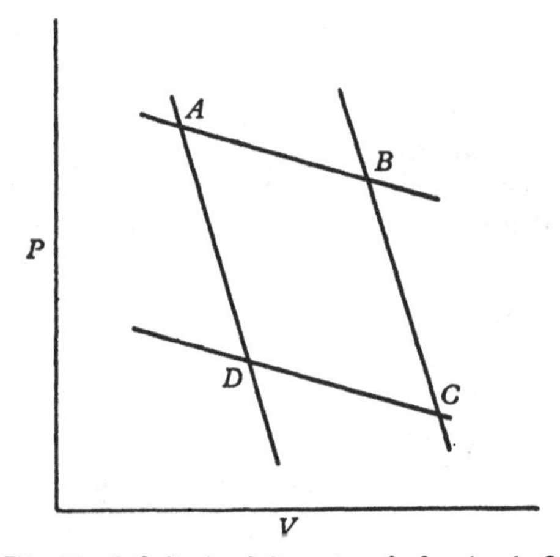

Fig. 11. Infinitesimal Carnot cycle for simple fluid.

infinitesimal portions of isotherms at temperatures T and $T - \delta T$, and BC and AD are adiabatics. Let the volume change between A and B be δV; the vertical distance between the two isotherms is $(\partial P/\partial T)_V\,\delta T$, so that

$$\delta A = \left(\frac{\partial P}{\partial T}\right)_V \delta V\,\delta T. \tag{5·13}$$

Now in the course of the reversible isothermal expansion AB the heat absorbed is $T(\partial S/\partial V)_T\,\delta V$. Therefore from (5·11) and (5·13),

$$\frac{(\partial P/\partial T)_V\,\delta V\,\delta T}{T(\partial S/\partial V)_T\,\delta V} = \frac{\delta T}{T},$$

i.e.

$$\left(\frac{\partial P}{\partial T}\right)_V = \left(\frac{\partial S}{\partial V}\right)_T,$$

which is just M.4. We should not therefore expect this graphical method to yield any result which could not be derived, probably with less labour, by the direct application of a Maxwell relation. In fact, the Clapeyron equation may be derived immediately from M.4 if it is applied to a mixture of a vapour and liquid in equilibrium. For so long as both phases are present $(\partial P/\partial T)_V$ is independent of V and equal to the variation of vapour pressure with temperature. In consequence $(\partial S/\partial V)_T$ is also independent of V. Let us therefore make a finite expansion in which unit mass of liquid is evaporated; then $\Delta V = v_v - v_l$, and $\Delta S = l/T$. Substituting $\Delta S/\Delta V = (\partial S/\partial V)_T$ in M.4 we reach Clapeyron's equation at once.

It is possible that the reader may feel some uneasiness in applying Maxwell's relations, which were derived for a simple fluid, to a problem in which liquid and vapour coexist in an inhomogeneous mixture. If so, let him imagine the liquid and vapour to be enclosed in an opaque cylinder fitted with a piston, so that he may think himself to be dealing with a peculiar fluid whose isotherms have the unusual form of fig. 10. In fact, Maxwell's equations apply, not simply to a fluid, but to any system which is determined by two parameters, V and T say, and the mixture of liquid and vapour belongs to this class.

The functions U, H, F and G

We have introduced in this chapter the four functions, *internal energy* U, *enthalpy* H, *free energy* F and *Gibbs function* G, which are related to one another and to P, V, S and T by the definitions

$$H \equiv U + PV,$$

$$F \equiv U - TS,$$

$$G \equiv U - TS + PV \equiv F + PV \equiv H - TS.$$

The differential coefficients of these functions may be expressed in the following forms:

$$dU = T\,dS - P\,dV,$$

$$dH = T\,dS + V\,dP,$$

$$dF = -S\,dT - P\,dV,$$

$$dG = -S\,dT + V\,dP,$$

and it was by use of these forms that we derived Maxwell's equations. Since the functions play important roles in the application of thermodynamics to specific problems we shall outline some of their properties here.

First it will be observed that a specification of each as a function of the appropriate parameters provides a complete specification of the thermodynamic properties of the system. The appropriate parameters for each function are as follows: $U(S, V)$, $H(S, P)$, $F(T, V)$, $G(T, P)$. For example, if $U(S, V)$ is a known function, the temperature and pressure corresponding to any particular state (S, V) may be found immediately, since $(\partial U/\partial S)_V = T$ and $(\partial U/\partial V)_S = -P$. Hence any other of the thermodynamic functions may be constructed and in addition the equation of state is found. It is necessary for this purpose that each function shall be known in terms of the appropriate parameters and not any two parameters. If, for instance, we know $U(T, V)$ we cannot derive therefrom the equation of state. It is this circumstance which gives the free energy its particular importance, since the appropriate parameters are V and T. In many problems involving the calculation of the properties of a system by the use of statistical mechanics it is most convenient to consider the system as defined by its volume and temperature, and therefore it is only necessary to calculate the free energy in order to derive all the equilibrium properties, which are related to F by the equations

$$(\partial F/\partial T)_V = -S,$$

$$(\partial F/\partial V)_T = -P,$$

$$U = F - T(\partial F/\partial T)_V = -T^2\left(\frac{\partial}{\partial T}\frac{F}{T}\right)_V. \tag{5·14}$$

Equation (5·14) is known as the Gibbs-Helmholtz equation. Analogous results may be similarly derived:

$$G = -V^2\left(\frac{\partial}{\partial V}\frac{F}{V}\right)_T,$$

$$H = -T^2\left(\frac{\partial}{\partial T}\frac{G}{T}\right)_P.$$

In any reversible change taking place at constant temperature,

$$\begin{aligned}\Delta F &= \Delta U - T\Delta S\\ &= \Delta U - Q\\ &= W, \text{ from the first law.}\end{aligned}$$

From this it follows that the work done by a system in an isothermal reversible change is equal to its decrease in free energy. This is the origin of the name ascribed to F. It will be seen from the arguments to be presented in Chapter 7 that the decrease in F of a system represents the *maximum* work which can be performed as a consequence of a given isothermal change, and that this maximum is only achieved by carrying out the change reversibly.

CHAPTER 6

APPLICATIONS OF THERMODYNAMICS TO SIMPLE SYSTEMS

Introduction

We shall see in this chapter how thermodynamics may be used to correlate the thermal properties of simple systems, such as fluids, in the equilibrium state. Taking fluids (or solids subjected only to hydrostatic pressure) as typical of two-parameter systems we may tabulate the thermal properties† as follows:

(1) The equation of state. This includes, for liquids and solids, such properties as isothermal compressibility and expansion coefficient.

(2) The adiabatic equation, the relation between any two of P, V and T in an adiabatic change.

(3) The specific heats, particularly the two principal specific heats C_P, the specific heat at constant pressure and C_V, the specific heat at constant volume. In general these are not constant but are functions of the parameters of state.

(4) The Joule coefficient, the relation between temperature and volume when a fluid (especially a gas) is expanded from one equilibrium state to another in an isolated enclosure without doing work.

(5) The Joule-Kelvin coefficient, the relation between temperature and pressure when a fluid (especially a gas) is expanded from one equilibrium state to another through a throttle valve.

If all these properties had to be measured individually for all states of the fluid the amount of experimental effort required would be enormous, but fortunately they may be so closely related by thermodynamic reasoning that the number of independent properties which must be measured in order to give complete information is not too formidable. The properties which may be determined most easily vary according to the system studied, but usually they are the equation of state and one of the principal specific heats, for solids and liquids C_P and for gases either C_P or C_V. Specific heats are most accurately determined over a wide range of temperature by means of an adiabatic (Nernst) calorimeter, and this will provide values of C_P at zero pressure for solids fairly readily and values of C_V at moderate pressures for

† Non-equilibrium thermal properties, such as thermal conductivity, are of course beyond the range of classical thermodynamics, and the Joule and Joule-Kelvin effects tabulated below ((4) and (5)) are amenable to thermodynamic methods only because the initial and final states are equilibrium states.

gases with somewhat more trouble; the specific heat which is measured for a liquid will depend on whether the liquid is given room to expand or not, but there is not usually much difficulty in deciding what is actually measured. Flow methods for determining C_P for a gas have been developed to give results of high accuracy. The other properties, except perhaps the Joule-Kelvin coefficient, present considerable difficulties in their accurate determination. There are occasions, as will be mentioned in due course, when measurements of the Joule-Kelvin coefficient may usefully supplement studies of the equation of state.

It will be seen, then, that the amount of experimental information which may be obtained by standard means is rather limited, but fortunately it is sufficient, as we may easily show. All the properties listed above may be formulated in terms of thermodynamic parameters:

(1) The equation of state expresses the relation between P, V and T in equilibrium,

$$f(P, V, T) = 0.$$

(2) An adiabatic change is one which takes place at constant entropy, so that complete knowledge of the differential coefficient $(\partial P/\partial V)_S$ for all states is sufficient to determine the relation between P and V in an adiabatic change; similarly, knowledge of $(\partial P/\partial T)_S$ and $(\partial V/\partial T)_S$ determines the relations between P and T, and V and T, respectively.

(3) The specific heat is defined as $\mathrm{d}Q/\mathrm{d}T$ in a reversible change under specified conditions. According to the first law,

$$\mathrm{d}Q = \mathrm{d}U + P\,\mathrm{d}V;$$

therefore

$$C_V = (\partial U/\partial T)_V \tag{6·1}$$

and

$$C_P = (\partial U/\partial T)_P + P(\partial V/\partial T)_P. \tag{6·2}$$

Alternatively, since in a reversible change $\mathrm{d}Q = T\,\mathrm{d}S$,

$$C_V = T(\partial S/\partial T)_V \tag{6·3}$$

and

$$C_P = T(\partial S/\partial T)_P. \tag{6·4}$$

Yet another convenient expression for C_P may be obtained in terms of the enthalpy H. For $\mathrm{d}H = \mathrm{d}U + P\,\mathrm{d}V + V\,\mathrm{d}P$ in general, so that $\mathrm{d}H = \mathrm{d}Q + V\,\mathrm{d}P$ for a reversible change, and

$$C_P = (\partial H/\partial T)_P. \tag{6·5}$$

(4) In the expansion of an isolated fluid without the performance of work, the internal energy stays constant, since $Q = W = 0$. The Joule coefficient is therefore, as we shall discuss in more detail later, $(\partial T/\partial V)_U$.

(5) We shall see that in a throttling process the enthalpy is conserved, and the Joule-Kelvin coefficient is given by $(\partial T/\partial P)_H$.

If, therefore, we know all the thermodynamic functions for all states of the fluid we can derive complete information. But the thermodynamic functions themselves are not independent; if we know the equation of state and the value of any of U, S, H (or F or G for that matter) for all states of the fluid, the values of the others may be derived. If, for example, we have a complete specification of $S(T, V)$ we may use the fundamental equation

$$\mathrm{d}U = T\,\mathrm{d}S - P\,\mathrm{d}V$$

to find $\mathrm{d}U$ for any infinitesimal change $(\mathrm{d}T, \mathrm{d}V)$; for the equation of state gives the value of P at any (T, V) and the rest of the quantities on the right-hand side are known or defined by the change considered. The function U may therefore in principle be constructed by step-by-step integration. A similar result may easily be shown to hold for any other thermodynamic function.

Lastly, we may note that it is not necessary to know by experiment the value, say, of $S(T, V)$ for all values of T and V, since a little experimental information may be largely extended by use of thermodynamics and the equation of state. Since S is here treated as a function of T and V, we have that

$$\mathrm{d}S = \left(\frac{\partial S}{\partial T}\right)_V \mathrm{d}T + \left(\frac{\partial S}{\partial V}\right)_T \mathrm{d}V$$

$$= \frac{C_V}{T}\,\mathrm{d}T + \left(\frac{\partial P}{\partial T}\right)_V \mathrm{d}V, \quad \text{by use of M.4.}$$

Hence $$S(T_1, V_1) - S(T_0, V_0) = \int_{T_0, V_0}^{T_1, V_0} \frac{C_V}{T}\,\mathrm{d}T + \int_{T_1, V_0}^{T_1, V_1} \left(\frac{\partial P}{\partial T}\right)_V \mathrm{d}V. \qquad (6\cdot 6)$$

The integrations are supposed here to be performed in two stages, the first integral being evaluated at constant volume from (T_0, V_0) to (T_1, V_0), and then the second evaluated at constant temperature from (T_1, V_0) to (T_1, V_1). Since knowledge of the first integrand demands knowledge only of C_V as a function of temperature at a certain standard volume V_0, and the second integrand depends on the equation of state, it is clear that this is all the information needed to determine S completely, and hence to find all the listed properties for all states of the fluid. We have taken a special case to illustrate this point, but the reader may verify that similar results hold for the other thermodynamic variables, and that a complete specification is given by the following data:

(1) The equation of state.

(2) One of the specific heats along any line on the (P, T) or (V, T)

diagram, provided that the line runs through all temperatures for which the information is needed.

Having thus demonstrated how little experimental information is needed, we now turn our attention to the explicit evaluation of the listed properties in terms of the measurable quantities, beginning with the specific heats.

Relations between the specific heats

Since we have seen that there is no need to measure the specific heats in all states of the fluid, but only to have knowledge of one specific heat along one line on the (P, T) or (V, T) diagram, we expect to be able to deduce C_P from C_V or vice versa, and also to determine either at a given temperature for all pressures or volumes, once a single value is found experimentally. For this purpose it is convenient to express $C_P - C_V$, $(\partial C_P/\partial P)_T$ and $(\partial C_V/\partial V)_T$ in terms of quantities which can be calculated from the equation of state.

Let us first consider $(\partial C_P/\partial P)_T$ and $(\partial C_V/\partial V)_T$. Since from (6·4) $C_P = T(\partial S/\partial T)_P$, we have that

$$\left(\frac{\partial C_P}{\partial P}\right)_T = T\left(\frac{\partial}{\partial P}\right)_T\left(\frac{\partial S}{\partial T}\right)_P = T\left(\frac{\partial}{\partial T}\right)_P\left(\frac{\partial S}{\partial P}\right)_T = -T\left(\frac{\partial^2 V}{\partial T^2}\right)_P \quad \text{from M.3.} \tag{6·7}$$

Similarly,

$$\left(\frac{\partial C_V}{\partial V}\right)_T = T\left(\frac{\partial^2 P}{\partial T^2}\right)_V \quad \text{by use of M.4.} \tag{6·8}$$

The quantities on the right-hand sides of (6·7) and (6·8) may be evaluated from the equation of state. The reader should verify that they both vanish for a perfect gas (as is to be expected since U and H are functions of temperature only) and that while $(\partial C_V/\partial V)_T = 0$ for a van der Waals gas, $(\partial C_P/\partial P)_T \neq 0$.

In order to calculate $C_P - C_V$ it is convenient to regard S as a function of T and V, so that

$$\mathrm{d}S = \left(\frac{\partial S}{\partial T}\right)_V \mathrm{d}T + \left(\frac{\partial S}{\partial V}\right)_T \mathrm{d}V,$$

or

$$\left(\frac{\partial S}{\partial T}\right)_P = \left(\frac{\partial S}{\partial T}\right)_V + \left(\frac{\partial S}{\partial V}\right)_T\left(\frac{\partial V}{\partial T}\right)_P.$$

Hence, from (6·3) and (6·4),

$$C_P - C_V = T\left(\frac{\partial S}{\partial V}\right)_T\left(\frac{\partial V}{\partial T}\right)_P$$

$$= T\left(\frac{\partial P}{\partial T}\right)_V\left(\frac{\partial V}{\partial T}\right)_P \quad \text{from M.4.} \tag{6·9}$$

This expression for $C_P - C_V$ is convenient if the equation of state is known explicitly, but for solids and liquids it is usually more convenient to express the result in terms of the expansion coefficient β $\left(\equiv \frac{1}{V}\left(\frac{\partial V}{\partial T}\right)_P\right)$ and the isothermal compressibility $k_T\left(\equiv -\frac{1}{V}\left(\frac{\partial V}{\partial P}\right)_T\right)$, which is the reciprocal of the isothermal bulk modulus. Since, by (5·8),

$$\left(\frac{\partial P}{\partial T}\right)_V = -\left(\frac{\partial P}{\partial V}\right)_T\left(\frac{\partial V}{\partial T}\right)_P,$$

(6·9) can be put in the form

$$C_P - C_V = -T\left(\frac{\partial V}{\partial T}\right)_P^2\left(\frac{\partial P}{\partial V}\right)_T$$

$$= VT\beta^2/k_T. \tag{6·10}$$

This form of the result shows that C_P can never be less than C_V, since k_T is always positive; $C_P = C_V$ when $\beta = 0$, as, for example, in water at 4° C. or liquid helium at about 1° K.

Equations (6·7), (6·8) and (6·9) or (6·10) provide all the thermodynamic information needed to extend the bare minimum of data concerning the specific heat into complete knowledge for all states of the fluid. In addition, the ratio of the principal specific heats $\gamma\ (\equiv C_P/C_V)$ is of course determined by either (6·9) or (6·10),

$$\gamma - 1 = \frac{T}{C_V}\left(\frac{\partial P}{\partial T}\right)_V\left(\frac{\partial V}{\partial T}\right)_P = \frac{VT\beta^2}{C_V k_T}. \tag{6·11}$$

In general γ varies with the state of the fluid, even for a perfect gas if, as is quite possible, C_V depends on temperature (as in hydrogen). The physical importance of γ is twofold: in the first place its value for a gas provides evidence concerning the number of degrees of freedom of the molecules constituting the gas, a matter which has no place in our present discussion, and in the second place the adiabatic equation of a gas is conveniently expressed in terms of γ.

The adiabatic equation

If we wish to find the relationship between P and V in an adiabatic change, we naturally seek to find an expression for $(\partial P/\partial V)_S$ in terms of measured properties of the fluid (or solid if only hydrostatic stresses are applied), in which category we are now able to include γ. This is very simply done, for

$$\frac{(\partial P/\partial V)_S}{(\partial P/\partial V)_T} = \frac{(\partial S/\partial V)_P(\partial P/\partial S)_V}{(\partial T/\partial V)_P(\partial P/\partial T)_V} = \frac{(\partial S/\partial T)_P}{(\partial S/\partial T)_V} = \gamma, \tag{6·12}$$

by use of the mathematical identities (5·8) and (5·9). Thus the adiabatic bulk modulus is γ times the isothermal bulk modulus for all fluids. Equation (6·12) expresses the general solution of the problem of determining the adiabatic relationship between P and V, for $(\partial P/\partial V)_T$ may be calculated from the equation of state and the adiabatic curve is found by integrating the equation

$$\left(\frac{\partial P}{\partial V}\right)_S = \gamma \left(\frac{\partial P}{\partial V}\right)_T, \tag{6·13}$$

if necessary step-by-step. For example, if we consider a perfect gas, $PV = RT$ and $(\partial P/\partial V)_T = -P/V$, so that (6·13) takes the form

$$\left(\frac{\partial P}{\partial V}\right)_S = -\gamma P/V,$$

with solution $PV^\gamma =$ constant, if γ is constant.

The fact that we have derived (6·12) by using only mathematical identities, and without employing Maxwell's relations, suggests that the result does not depend on the second law, and this is in fact so. We shall now show how the first law alone leads to the same result. For this purpose consider an infinitesimal portion of the indicator diagram, as in fig. 12, so that the two neighbouring isotherms and the adiabatic shown may be represented by straight lines; moreover, the internal energy may be considered to vary in a linear manner over the range shown; if the point A is (P_0, V_0),

$$U(P_0 + \delta P, V_0 + \delta V) = U(P_0, V_0) + \left(\frac{\partial U}{\partial P}\right)_V \delta P + \left(\frac{\partial U}{\partial V}\right)_P \delta V,$$

higher order terms in the expansion being neglected. We shall now calculate the difference in U between A and C, first along the path ABC and then round the path ADC. Equating these two gives the desired result. For the path ABC we have that

$$U_B - U_A = C_V \delta T,$$

and therefore
$$U_C - U_A = \frac{CA}{BA} C_V \delta T = g C_V \delta T, \tag{6·14}$$

where g is the ratio of the gradients of CD and BD, i.e.

$$(\partial P/\partial V)_{\text{adiabatic}}/(\partial P/\partial V)_{\text{isothermal}}.$$

For the path ADC we have that

$$U_D - U_A = C_P \delta T - P_0 \delta V, \quad \text{where} \quad \delta V = AD,$$

and $U_C - U_D = P_0 \delta V$, since DC is an adiabatic. Therefore

$$U_C - U_A = C_P \delta T. \tag{6·15}$$

Hence from (6·14) and (6·15),

$$gC_V = C_P \quad \text{or} \quad g = \gamma,$$

which is equivalent to (6·12).

By retaining γ in the expression for $(\partial P/\partial V)_S$ we avoid recourse to the second law; but if we do not know γ, but only C_P or C_V, we must

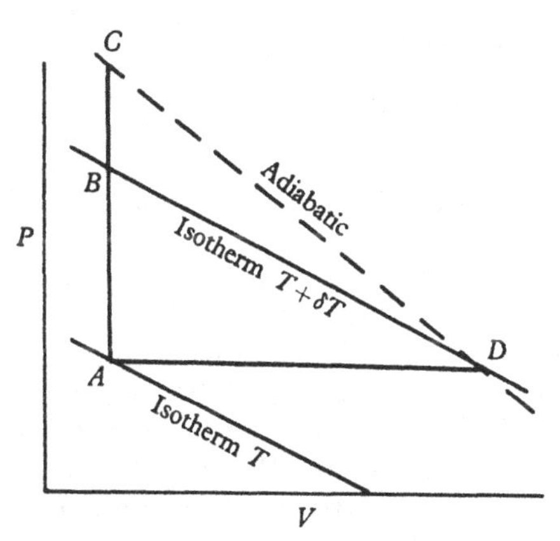

Fig. 12. Indicator diagram for simple fluid.

invoke the second law through (6·11). So too for the other adiabatic coefficients; thus

$$\left(\frac{\partial T}{\partial P}\right)_S = -\left(\frac{\partial T}{\partial S}\right)_P \left(\frac{\partial S}{\partial P}\right)_T \quad \text{from (5·8)}$$

$$= \frac{T}{C_P}\left(\frac{\partial V}{\partial T}\right)_P \quad \text{from M.3.} \tag{6·16}$$

Similarly,

$$\left(\frac{\partial T}{\partial V}\right)_S = -\frac{T}{C_V}\left(\frac{\partial P}{\partial T}\right)_V. \tag{6·17}$$

Magnetic analogues of the foregoing results

Let us now apply to magnetizable bodies the principles which we have used in deriving the properties of fluids, taking as the fundamental equation

$$dU' = T\,dS - P\,dV + \mathscr{H}.\,d\mathscr{M}. \tag{6·18}$$

To simplify the notation we shall replace the scalar product $\mathscr{H}.\,d\mathscr{M}$ by $\mathscr{H}\,d\mathscr{M}$, in which $\mathscr{M}$ is to be interpreted as the component of $\mathscr{M}$ in the direction of $\mathscr{H}$. We shall also, as (6·18) implies, only consider cases in which the external field is uniform. Under many circumstances it is permissible to neglect the term $P\,\mathrm{d}V$ in comparison with $\mathscr{H}\,\mathrm{d}\mathscr{M}$, and we shall first examine the relative magnitudes of these quantities under different conditions. It is convenient to note that a large number of analogues of Maxwell's relations may be derived from (6·18), of which the particular one we require for the present purpose is obtained by considering the differential properties of the Gibbs function G' ($\equiv U'-TS+PV$):

$$\mathrm{d}G' = -S\,\mathrm{d}T + V\,\mathrm{d}P + \mathscr{H}\,\mathrm{d}\mathscr{M};$$

therefore
$$\left(\frac{\partial G'}{\partial P}\right)_{T,\mathscr{M}} = V \quad \text{and} \quad \left(\frac{\partial G'}{\partial \mathscr{M}}\right)_{T,P} = \mathscr{H},$$

whence
$$\left(\frac{\partial V}{\partial \mathscr{M}}\right)_{T,P} = \left(\frac{\partial \mathscr{H}}{\partial P}\right)_{T,\mathscr{M}}. \tag{6·19}$$

The remaining twenty-three analogues may be derived by similar methods. Now let us imagine a differential change taking place at constant temperature and pressure, in which a variation of $\mathscr{H}$ leads to a change in both V and $\mathscr{M}$; clearly the ratio, $r_{T,P}$, of the two terms in (6·18), $P\,\mathrm{d}V$ and $\mathscr{H}\,\mathrm{d}\mathscr{M}$, is given by the expression

$$\begin{aligned} r_{T,P} &= \frac{P}{\mathscr{H}}\left(\frac{\partial V}{\partial \mathscr{M}}\right)_{T,P} \\ &= \frac{P}{\mathscr{H}}\left(\frac{\partial \mathscr{H}}{\partial P}\right)_{T,\mathscr{M}} \quad \text{from (6·19).} \end{aligned} \tag{6·20}$$

For the purpose of estimating the magnitude of $r_{T,P}$ let us assume that the body is linearly magnetizable, i.e. $\mathscr{M} = \alpha(T,P)\mathscr{H}$, where α is the susceptibility of the body itself. Then from (6·20),

$$r_{T,P} = -\frac{P}{\alpha}\left(\frac{\partial \alpha}{\partial P}\right)_T = Pk_T\frac{V}{\alpha}\left(\frac{\partial \alpha}{\partial V}\right)_T,$$

where k_T is the isothermal compressibility. For many magnetic substances (particularly paramagnetic substances), the property of magnetizability owes its origin to the internal structure of one of the atoms composing the substance, and under these conditions α is to all intents and purposes independent of the pressure or volume, since each atom acts nearly independently. For a paramagnetic gas, such as oxygen, $Pk_T = 1$, but $\frac{V}{\alpha}\left(\frac{\partial \alpha}{\partial V}\right)_T \ll 1$, so that $r_{T,P} \ll 1$; for a solid $\frac{V}{\alpha}\left(\frac{\partial \alpha}{\partial V}\right)_T$ may not be so small as for a gas, but it is unlikely to exceed the order of magnitude of unity, while $Pk_T \sim 10^{-5}$ to 10^{-6} for most solids at atmospheric pressure, so that $r_{T,P}$ is only likely to be comparable with unity at extremely high pressures or with exceptional substances for which $\dfrac{V}{\alpha}\left(\dfrac{\partial \alpha}{\partial V}\right)_T \gg 1$. On the other hand, if conditions are such that T and $\mathscr{H}$, or P and $\mathscr{H}$, are kept constant it is not so clear that the term $P\,\mathrm{d}V$ is negligible; indeed, if in the former case α is independent of pressure, or in the latter case α is independent of temperature, then $\mathscr{H}\,\mathrm{d}\mathscr{M}=0$ and the only work term is that due to pressure. It would therefore be necessary in applying thermodynamics under these conditions to examine the relative magnitudes of the terms carefully before discarding any. For present purposes, however, we shall be concerned only with phenomena in solids, which may be examined experimentally at very low pressures, and no significant error will result from assuming for the fundamental equation

$$\mathrm{d}U' = T\,\mathrm{d}S + \mathscr{H}\,\mathrm{d}\mathscr{M}. \tag{6·21}$$

All relations between the principal specific heats which were derived for fluids may now be taken over without further thought by replacing P and V by $\mathscr{H}$ and $-\mathscr{M}$. Thus

$$\left(\frac{\partial C_{\mathscr{H}}}{\partial \mathscr{H}}\right)_T = T\left(\frac{\partial^2 \mathscr{M}}{\partial T^2}\right)_{\mathscr{H}} = \mathscr{H} T\left(\frac{\partial^2 \alpha}{\partial T^2}\right)_{\mathscr{H}}, \tag{6·22}$$

in which α, the susceptibility $\mathscr{M}/\mathscr{H}$ of the body, need not be independent of $\mathscr{H}$;

$$\left(\frac{\partial C_{\mathscr{M}}}{\partial \mathscr{M}}\right)_T = -T\left(\frac{\partial^2 \mathscr{H}}{\partial T^2}\right)_{\mathscr{M}}; \tag{6·23}$$

$$C_{\mathscr{H}} - C_{\mathscr{M}} = -T\left(\frac{\partial \mathscr{H}}{\partial T}\right)_{\mathscr{M}}\left(\frac{\partial \mathscr{M}}{\partial T}\right)_{\mathscr{H}} = T\left(\frac{\partial \mathscr{M}}{\partial T}\right)_{\mathscr{H}}^2 \Big/ \left(\frac{\partial \mathscr{M}}{\partial \mathscr{H}}\right)_T. \tag{6·24}$$

Remembering that by definition $\mathscr{M}=\alpha(\mathscr{H},T)\mathscr{H}$, and defining the differential isothermal susceptibility α'_T as $(\partial \mathscr{M}/\partial \mathscr{H})_T$,† we can cast

† If the magnetization of the body is proportional to the field, as is commonly the case except at low temperatures and high field strengths (we exclude ferromagnetics from this discussion for the reasons explained in Chapter 3), then the generally defined susceptibility α ($\equiv \mathscr{M}/\mathscr{H}$), which may be a function both of T and $\mathscr{H}$, becomes simply a function of T and identical with the isothermal differential susceptibility α'_T.

It should be remembered in all this section that α is defined for a particular body, and is in general dependent upon the shape of the body. If we consider a long rod set parallel to $\mathscr{H}$, the field $\mathscr{H}_i$ within the body is simply $\mathscr{H}$, and α is the same as $V\chi$, where χ is the volume susceptibility, $\mathscr{I}/\mathscr{H}_i$, of the material. For any other shape, however, $\mathscr{H}_i$ and $\mathscr{H}$ are not the same. The only shapes which may be handled easily are ellipsoids, for which $\mathscr{H}_i$ is uniform. If one of the axes of the ellipsoids coincides in direction with $\mathscr{H}$, then $\mathscr{H}_i$ is also parallel to

(6·24) into the convenient form

$$C_{\mathscr{H}} - C_{\mathscr{M}} = \frac{T\mathscr{H}^2}{\alpha'_T}\left(\frac{\partial\alpha}{\partial T}\right)^2. \tag{6·25}$$

From this expression we see that when $\mathscr{H}=0$, $C_{\mathscr{H}}=C_{\mathscr{M}}$, as would be expected, and that for a paramagnetic material having α'_T positive $C_{\mathscr{H}} \geqslant C_{\mathscr{M}}$, while for a diamagnetic material having α'_T negative $C_{\mathscr{H}} \leqslant C_{\mathscr{M}}$.

By analogy with α'_T we may also define a differential adiabatic susceptibility α'_S as $(\partial\mathscr{M}/\partial\mathscr{H})_S$, the relationship between α'_T and α'_S being analogous to that between the compressibilities k_T and k_S of a fluid,

$$\alpha'_T = \Gamma\alpha'_S, \quad \text{in which} \quad \Gamma = C_{\mathscr{H}}/C_{\mathscr{M}}. \tag{6·26}$$

The adiabatic susceptibility of a magnetic material is what is usually measured by any method which involves alternating fields (as, for instance, measuring in an a.c. bridge the inductance of a coil containing the sample as core), since there is usually no time for the sample to exchange heat with its surroundings. From (6·25) it will be seen that $\Gamma=1$ when $\mathscr{H}=0$, and departs from unity as $\mathscr{H}^2$, so that when small measuring fields are used no error results from the assumption that it is α'_T which is measured. But if a large steady field be applied while the a.c. measurements are made there may be a significant difference between α'_T and α'_S. The implication of such a difference is that when a body is magnetized adiabatically it may change its temperature, and this is seen to be so by writing down the other adiabatic equations, the analogues of (6·16) and (6·17),

$$\left(\frac{\partial T}{\partial\mathscr{H}}\right)_S = -\frac{\mathscr{H}T}{C_{\mathscr{H}}}\left(\frac{\partial\alpha}{\partial T}\right)_{\mathscr{H}}, \tag{6·27}$$

$$\left(\frac{\partial T}{\partial\mathscr{M}}\right)_S = -\frac{\mathscr{H}T}{\alpha C_{\mathscr{M}}}\left(\frac{\partial\alpha}{\partial T}\right)_{\mathscr{M}}. \tag{6·28}$$

$\mathscr{H}$, and may conveniently be connected with $\mathscr{H}$ by introducing the idea of a *demagnetizing coefficient*, n, such that

$$\mathscr{H}_i = \mathscr{H} - 4\pi n\mathscr{I}.$$

The values of n range from zero (long rod parallel to $\mathscr{H}$) to unity (flat slab normal to $\mathscr{H}$), with such typical intermediate values as $\frac{1}{3}$ for a sphere and $\frac{1}{2}$ for a long circular cylinder normal to $\mathscr{H}$. The relation between α and χ is now readily derived:

$$\alpha = \frac{V\chi}{1+4\pi n\chi}.$$

If, as in many practical applications, $\chi < 10^{-4}$, the demagnetizing correction is very small and α may be taken as $V\chi$ without serious error, whatever the shape of the body. But in some cases of interest, particularly paramagnetic salts at low temperatures for which χ may approach unity, and superconductors for which $4\pi\chi$ is -1, the shape of the body may play an important role in determining its behaviour.

From (6·27) it is seen that if α is temperature-dependent, there is indeed a change in temperature when a body is magnetized adiabatically. This is the *magneto-caloric effect*; it is of extremely small magnitude at room temperature, but may under favourable circumstances become large at low temperatures, where it is used for producing temperatures well below 1° K. by *adiabatic demagnetization*. The magnitude of the effect may be estimated by considering an idealized paramagnetic body which obeys Curie's law at all temperatures,

$$\chi = a/T, \quad \text{where } a \text{ is a constant.} \tag{6·29}$$

For such a body the entropy $S(\mathscr{H}, T)$ in a field $\mathscr{H}$ may be seen, by application of the analogue of M.3, to vary with $\mathscr{H}$ according to the equation

$$(\partial S/\partial \mathscr{H})_T = V\mathscr{H}\,\mathrm{d}\chi/\mathrm{d}T = -aV\mathscr{H}/T^2,$$

so that

$$S(\mathscr{H}, T) = S(0, T) - \tfrac{1}{2}aV\mathscr{H}^2/T^2, \tag{6·30}$$

where $S(0, T)$ is the entropy in zero field. If a process of adiabatic demagnetization is carried out, starting at temperature T_1 in a field $\mathscr{H}$, and ending at temperature T_2 in zero field, then according to (6·30) T_1 and T_2 are related by the equation

$$S(0, T_1) - \tfrac{1}{2}aV\mathscr{H}^2/T_1^2 = S(0, T_2). \tag{6·31}$$

Now at room temperature, on account of the presence of T_1^2 in its denominator, the second term in (6·31) is small for any achievable field $\mathscr{H}$; moreover, the specific heat in zero field is large, so that $S(0, T)$ is a steep function of T. This means that $T_1 - T_2$ is small, and approximately, from (6·31),

$$T_1 - T_2 = \frac{aV\mathscr{H}^2}{2TC_0}, \tag{6·32}$$

where C_0 is the value of $C_{\mathscr{H}}$ in zero field. Equation (6·32) can of course be deduced immediately from (6·27) on the assumption that $C_{\mathscr{H}}$ is independent of $\mathscr{H}$. For a typical paramagnetic salt at room temperature, with $\mathscr{H}$ equal to 10^4 gauss, $T_1 - T_2$ is less than 10^{-4} degrees. But as the temperature is lowered the denominator in (6·32) falls more rapidly than T, since, except over a certain range at a very low temperature, C_0 normally decreases steadily with lowering of the temperature, and the cooling effect becomes more pronounced. At temperatures of a few degrees absolute the approximations made in deriving (6·32) are invalid and a different approximation will serve to exhibit the sort of behaviour to be expected. As may be seen from (6·30)

$$C_{\mathscr{H}} = C_0 + aV\mathscr{H}^2/T^2, \tag{6·33}$$

and at low temperatures quite a moderate value of $\mathscr{H}$ will cause the second term to dominate the first. If then we neglect C_0, and combine (6·33) with (6·27), putting $\alpha = V\chi$, we arrive at the equation

$$\left(\frac{\partial T}{\partial \mathscr{H}}\right)_S = \frac{T}{\mathscr{H}},$$

or

$$T/\mathscr{H} = \text{constant}.$$

Thus the temperature falls in proportion to the magnetic field during adiabatic demagnetization, and a very considerable reduction in temperature is possible. It should be realized that this argument is based on an idealized model which neglects both saturation effects and departures from Curie's law (6·29); in addition, it is applicable only at such field strengths that C_0 in (6·33) may be neglected. All these effects, as might be expected, serve to prevent the absolute zero from being attained by simply reducing the field to zero. Nevertheless, by choosing a paramagnetic salt which obeys Curie's law to as low as possible a temperature, such as chrome alum or copper potassium sulphate, and by demagnetizing from a field of perhaps 20,000 gauss and an initial temperature of 1° K., temperatures as low as one or two thousandths of a degree may be reached, far below the limit of any other cooling method. To enter further into the rich field of thermodynamics and low-temperature research afforded by adiabatic demagnetization would take us too far beyond the scope of this book, and the reader is referred elsewhere for more detailed information.† We shall, however, return to the topic at the end of the present chapter in discussing methods of establishing the absolute scale of temperature.

The Joule and Joule-Kelvin effects

We now revert to the problem of expressing the thermal properties of fluids in terms of measurable quantities. So far we have considered only reversible changes between equilibrium states; the Joule and Joule-Kelvin effects are by contrast concerned with changes between equilibrium states effected by irreversible processes. In the Joule effect a fluid (particularly a gas) is allowed to expand into a vacuum, so that it increases its volume irreversibly without performing any work; in addition, the system is thermally isolated so that no heat enters or leaves. Under these conditions it clearly changes irreversibly from one equilibrium state to another having the same value of U. Now although the change certainly does not proceed by way of intermediate equilibrium states, we may assume for the purpose of calculating the effect that it does so. For the final state is specified precisely by the values of V and U, and it does not matter how the final state is

† C. G. B. Garrett, *Magnetic Cooling* (Harvard University Press, 1954).

attained since we are concerned solely with functions of state. Thus the temperature change accompanying an infinitesimal Joule expansion (the *Joule coefficient* of the fluid) may be ascertained by expressing $(\partial T/\partial V)_U$ in terms of known quantities, in the following way:

$$\left(\frac{\partial T}{\partial V}\right)_U = -\frac{(\partial U/\partial V)_T}{(\partial U/\partial T)_V} \quad \text{by (5·8)}$$

$$= -\frac{1}{C_V}\left\{T\left(\frac{\partial S}{\partial V}\right)_T - P\right\} \quad \text{from (5·2)}$$

$$= -\frac{1}{C_V}\left\{T\left(\frac{\partial P}{\partial T}\right)_V - P\right\} \quad \text{from M.4,} \qquad (6·34)$$

and this expresses the Joule coefficient in terms of C_V and quantities determinable from the equation of state. It is easily verified that the Joule coefficient vanishes for a perfect gas, as may be seen directly from the definition of a perfect gas as one for which U is a function of temperature only. The reader should evaluate the Joule coefficient for a van der Waals gas and attempt to understand from an atomic standpoint why it takes the form it does. For a finite volume change the resultant temperature change may of course be calculated in principle by integration of (6·34). There may, however, be a considerable gulf between principle and practice, for it is only with especially simple equations of state and variations of C_V with V and T that (6·34) can be cast into the form of a differential equation in V and T which may be solved with any ease. In general, it is probably less troublesome to return to first principles, and use available data to construct lines of constant U on a V-T diagram by processes analogous to those explained on p. 59 in connexion with the calculation of S. From such a diagram the magnitude of the cooling in a given Joule expansion is of course immediately plain.

The irreversibility of the Joule effect gives rise to a change in the entropy of the fluid even though the system is adiabatically enclosed. The change is readily calculated, since it follows immediately from (5·2) that $(\partial S/\partial V)_U = P/T$, as already remarked on p. 42. Therefore in the expansion the entropy always increases. For a perfect gas $P/T = R/V$ and the entropy change in a Joule expansion from V_1 to V_2 is $R \log (V_2/V_1)$.

The practical importance of the Joule effect is very small, but the same is not true of the related Joule-Kelvin effect, the temperature change which accompanies the expansion of a gas from high to low pressure through a throttle (in the early experiments on this phenomenon a porous plug was used instead of a mechanical valve, and the experiment is sometimes referred to as the porous-plug experiment).

The experimental arrangement is shown schematically in fig. 13; gas enters at a pressure P_0 and temperature T_0 and leaves at a lower pressure P_1 and, in general, a different temperature T_1. On each side of the throttle the gas is moving but is otherwise in thermal equilibrium. In the idealized form of the experiment it is assumed that there is no heat flow through the walls of the tubes.

This experimental arrangement is one example of a large class of *stationary flow* phenomena, which are characterized by a non-equilibrium state of a fluid system which maintains the same configuration from time to time. In its simplest form a system in stationary flow may consist simply of a tube of varying cross-section, as in fig. 13, through which a steady flow is maintained from external sources, and in which a constant pressure and temperature distribution is set up. But the tube need not be a real tube—in the streamline flow of a fluid it is

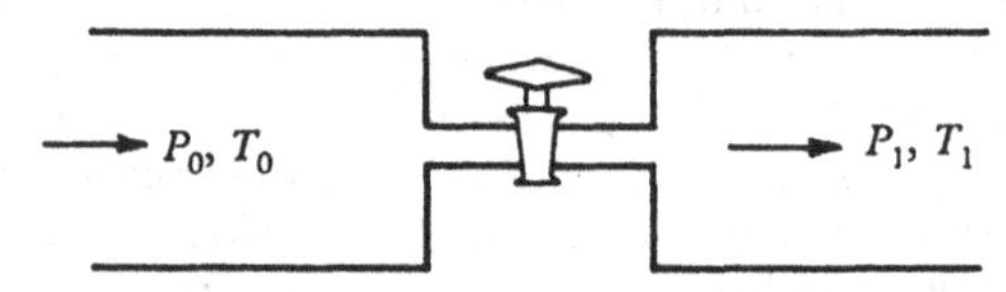

Fig. 13. Joule-Kelvin expansion through a throttle.

permissible to isolate for consideration any part of the fluid whose motion is bounded by streamlines. It is now easy to show from the first law that if there is no transport of heat by conduction either along or across the lines of flow the enthalpy H of a given portion of the fluid is conserved as it passes through the system. Consider the tube of flow in fig. 14, and let the elementary volumes described at each end contain the same mass, m, of fluid. Then as a quantity of fluid having mass m enters the shaded region at one end, an equal mass leaves at the other, and since the flow is stationary the internal energy of the fluid contained in the shaded region is unaltered. The factors tending to change this internal energy are:

the energy mu_0 transported in by the fluid entering (here and elsewhere we use small letters to denote the values of extensive thermodynamic quantities *per unit mass*);

the energy mu_1 transported out by the fluid leaving;

the work mP_0v_0 done by the fluid outside the shaded region on the fluid inside;

the work mP_1v_1 done by the fluid inside on the fluid outside.

If there is no heat contribution to the internal energy of the shaded region, we may therefore write

$$\left.\begin{aligned} u_0-u_1+P_0v_0-P_1v_1=0, \\ \text{or} \quad h_0=h_1, \quad \text{where } h \text{ is the enthalpy per unit mass.} \end{aligned}\right\} \tag{6·35}$$

Since the length of the shaded region was chosen arbitrarily, it follows that h is constant along a tube of flow. The result may obviously be generalized to cover more complicated stationary flows in which more

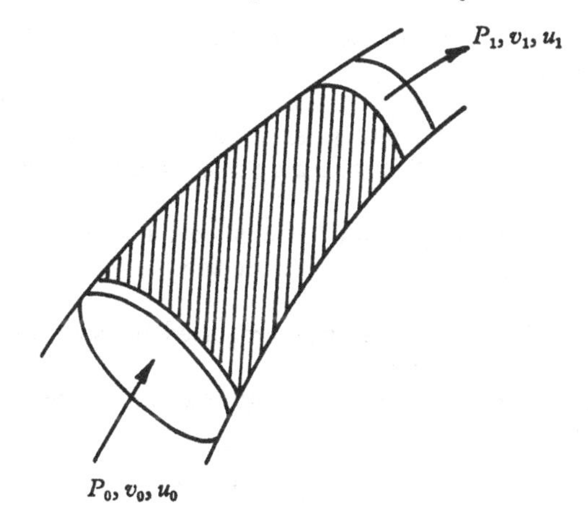

Fig. 14. Tube of flow, to illustrate Bernoulli's theorem.

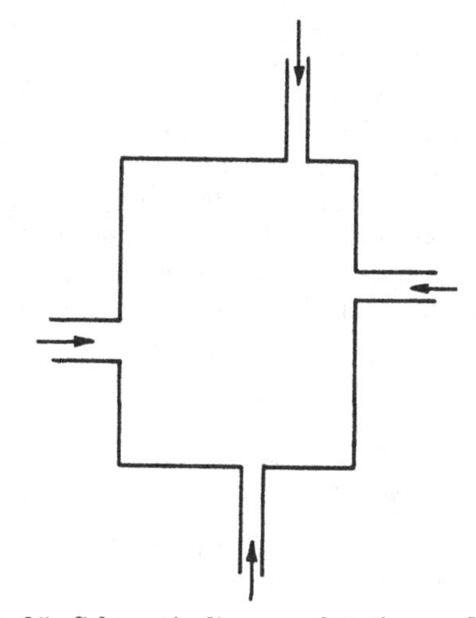

Fig. 15. Schematic diagram of stationary flow.

than a single stream is involved, as represented schematically in fig. 15. If the mass entering the system in unit time is m_i for the ith channel, and the enthalpy per unit mass is h_i, then $\sum_i m_i = 0$ and, provided there is no heat exchange between the system and its surroundings, and no heat flow along the channels at the points where the h_i are measured,

$$\sum_i m_i h_i = 0. \tag{6·36}$$

It should be noted that there need be no restriction on heat exchange within the system.

The constancy of h along a streamline is a result more familiarly known as Bernoulli's theorem when applied to an incompressible fluid. The internal energy density u may be considered as composed of a number of terms:

$u(0)$, the internal energy of unit mass at rest at a standard height, under the same conditions of temperature and pressure that obtain in the stream;

gx, the potential energy, the work required to raise the mass to a height x above the standard;

$\frac{1}{2}w^2$, the kinetic energy, the work required to accelerate the mass from rest to a velocity w.

Thus

$$h = u(0) + gx + \tfrac{1}{2}w^2 + P/\rho,$$

where ρ is the density. In particular, for an incompressible fluid the temperature and $u(0)$ remain constant along the stream if there is no heat transport, and therefore $gx + \frac{1}{2}w^2 + P/\rho$ is a constant of the flow, which is the well-known form of Bernoulli's theorem.

Let us now return to the Joule-Kelvin effect. If there is no heat flow at the output and input the enthalpy is conserved† during the irreversible pressure drop at the throttle, and if the flow is slow enough the kinetic energy may be neglected in calculating the enthalpy. We may use the same argument as with the Joule effect to replace this irreversible process by a hypothetical reversible process in which h is conserved, and therefore calculate the temperature change

$$T_1 - T_0 = \int_{P_0}^{P_1} \left(\frac{\partial T}{\partial P}\right)_h \mathrm{d}P. \tag{6·37}$$

We now express the *Joule-Kelvin coefficient* $(\partial T/\partial P)_h$ in terms of measurable quantities, as follows:

$$\left(\frac{\partial T}{\partial P}\right)_h = -\frac{(\partial h/\partial P)_T}{(\partial h/\partial T)_P}$$

$$= -\frac{1}{c_P}\left\{T\left(\frac{\partial s}{\partial P}\right)_T + v\right\},$$

since $\mathrm{d}h = T\,\mathrm{d}s + v\,\mathrm{d}P$ and $(\partial h/\partial T) = c_P$.

† Since this is an irreversible process the entropy need not be conserved. The reader should show that there is always an increase of entropy at the throttle.

Therefore, by use of M.3,

$$\left(\frac{\partial T}{\partial P}\right)_h = \frac{1}{c_P}\left\{T\left(\frac{\partial v}{\partial T}\right)_P - v\right\}, \tag{6·38}$$

which is the desired form, involving c_P and quantities calculable from the equation of state. The use of (6·38) in (6·37) to calculate the temperature change resulting from a finite pressure drop involves the same difficulties in general as were remarked on in connexion with the Joule effect (p. 69). Here also the practical treatment of the problem is to use experimental data to construct lines of constant enthalpy on a P-T diagram, or, as is obviously equivalent, isotherms on an H-P diagram, of which fig. 18 is an example.

It follows from (6·38) that the Joule-Kelvin effect vanishes for a perfect gas; this may be seen also from the fact that the definition of a perfect gas implies that h is a function of temperature only. It may also vanish for an imperfect gas under certain conditions, as may be seen by applying (6·38) to the equation of state expressed as a power series,

$$Pv = rT + BP + CP^2 + \dots, \tag{6·39}$$

in which r is the gas constant per unit mass and the *virial coefficients*† B, C, etc., are functions of temperature only. Combining this equation with (6·38), we find that

$$\left(\frac{\partial T}{\partial P}\right)_h = \frac{1}{c_P}\left\{\left(T\frac{\mathrm{d}B}{\mathrm{d}T} - B\right) + P\left(T\frac{\mathrm{d}C}{\mathrm{d}T} - C\right) + \dots\right\}. \tag{6·40}$$

For all gases the behaviour of B is qualitatively similar, as indicated in fig. 16. At not too high pressures the gas may be adequately represented by terminating the virial series after B. One can then see from fig. 16 that there is a temperature, the *Boyle temperature* T_B, at which $B=0$ and the gas approximates closely to a perfect gas, and that at a higher temperature, the *Inversion temperature* T_I, $T\mathrm{d}B/\mathrm{d}T - B$ vanishes and with it the Joule-Kelvin coefficient. Above the inversion temperature $\mathrm{d}B/\mathrm{d}T$ is less than B/T and the gas is warmed by the expansion; below T_I it is cooled. In showing inversion the Joule-Kelvin effect contrasts with the Joule effect, for real gases (in consequence of the fact that their intermolecular forces are always attractive at distances larger than the molecular diameter) always cool when expanded into a vacuum.

† Strictly this term should be reserved for the coefficients of the power series $Pv = A(1 + B/v + C/v^2 + \dots)$ introduced by Kamerlingh Onnes as an empirical equation of state; but we shall use it for similar equations in the absence of any alternative designation.

In order to examine the inversion phenomenon at higher pressures, further virial coefficients must be considered, or the equation of state must be known in a closed form. It is clear from (6·38) that the vanishing of the Joule-Kelvin coefficient occurs whenever $(\partial v/\partial T)_P = v/T$, and the substitution of this condition in the equation of state yields the inversion curve. We shall calculate the form of the inversion curve for a gas obeying Dieterici's equation,

$$P(v-b) = rT\,e^{-a/(rTv)}, \tag{6·41}$$

since the result is particularly simple to derive for this equation of state. To facilitate the interpretation of the result it is convenient to

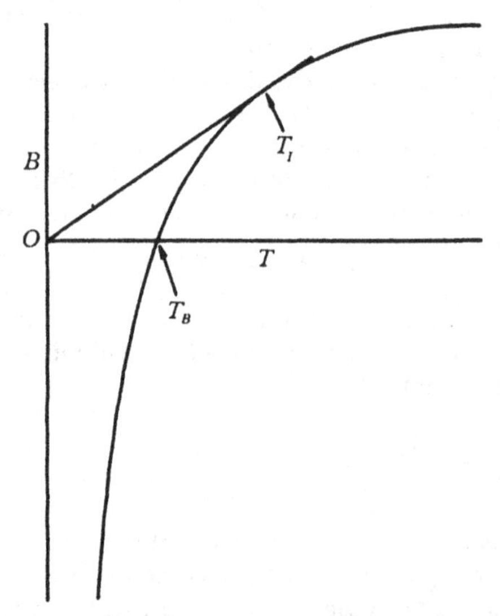

Fig. 16. Second virial coefficient, showing Boyle temperature (T_B) and inversion temperature (T_I).

use the equation in its reduced form ($\Pi = P/P_c$, $\Theta = T/T_c$, $\Phi = v/v_c$, where P_c, v_c and T_c are the critical pressure, volume and temperature, which may readily be shown to equal $a/(4e^2b^2)$, $2b$ and $a/(4br)$ respectively),

$$\Pi(2\Phi - 1) = \Theta \exp\left(2 - \frac{2}{\Theta\Phi}\right). \tag{6·42}$$

We now differentiate with respect to Θ, keeping Π constant, and replace $(\partial\Phi/\partial\Theta)_\Pi$ everywhere by Φ/Θ. Then after a little rearrangement, we find that the inversion curve on the Θ-Φ diagram takes the simple form

$$\Phi = \frac{4}{8-\Theta}. \tag{6·43}$$

In order to find the form of the inversion curve on the Θ-Π diagram we substitute (6·43) in (6·42), to yield the result, plotted in fig. 17,

$$\Pi = (8-\Theta)\exp\left(\frac{5}{2}-\frac{4}{\Theta}\right). \tag{6·44}$$

Comparison of this theoretical curve with a typical experimental curve, as given in fig. 17, shows that Dieterici's equation predicts the shape fairly closely, but is not so satisfactory as regards the scale; it is, however, quite as good as could be expected in view of the approximations

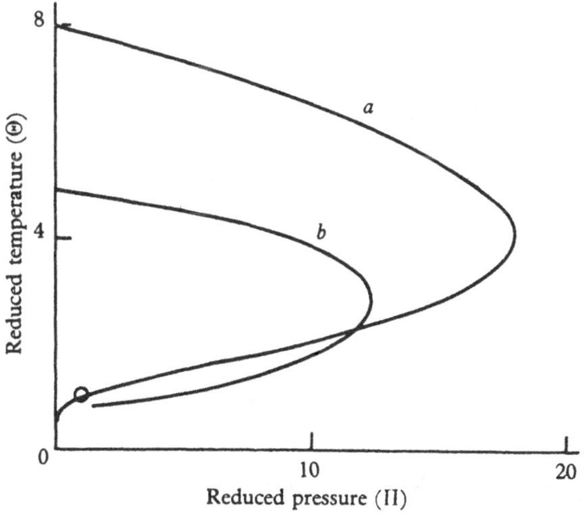

Fig. 17. Inversion curves: (*a*) according to Dieterici's equation; (*b*) experimental curve for nitrogen. The ringed point marks the critical point.

which are made in the derivation of Dieterici's equation.† Within the inversion curve a gas is cooled on expansion, outside it is warmed. The magnitude of the cooling which is obtainable may be seen from fig. 18, which shows the isotherms of helium on an H-P diagram. Joule-Kelvin expansion corresponds on this diagram to a horizontal movement to the left, so that the dotted locus of minima is the inversion curve. If helium at 10° K. and 20 atmospheres pressure is expanded to 1 atmosphere the process is that represented by the arrow, and the final temperature is about 6·5° K., or little over one-half of the initial temperature.

† The reader may find it instructive to derive the equation corresponding to (6·44) from van der Waals's equation, $(P+a/v^2)(v-b)=rT$; he will find (as applies to nearly all calculations based on these equations) that the derivation is not so straightforward, and the result when obtained not appreciably better.

There are a number of applications of Joule-Kelvin expansion which make it an important process, and the basic reason for its importance lies in the ease with which it may be carried out. Since it is a continuous process it is possible to make accurate determinations of the pressure drop and temperature change, and it is not difficult to arrange the gas circuits and throttle so that very nearly ideal conditions are realized as

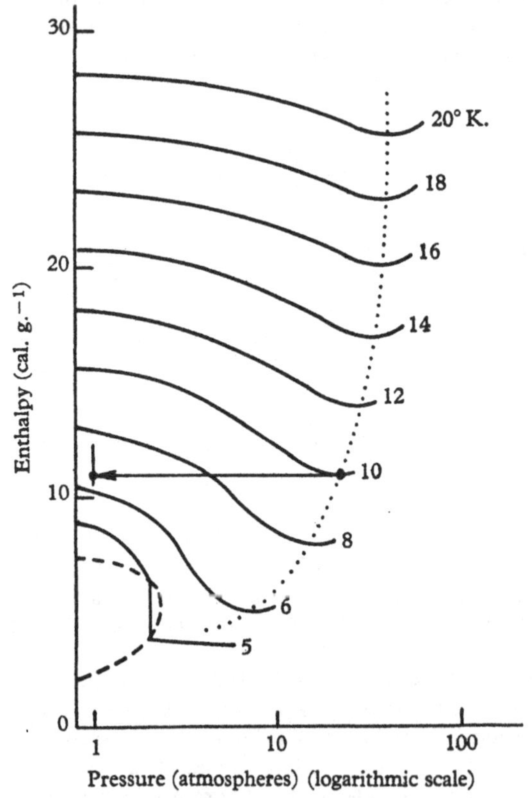

Fig. 18. Isotherms of helium on an *H-P* diagram (W. H. Keesom, *Helium*, p. 252, Elsevier, 1942).

regards heat insulation and gentle flow (to minimize the kinetic energy term in h).† It is therefore a convenient method for determining the isenthalps of a gas. Expansion from various pressures and temperatures determines the form of the isenthalps; measurement of c_p at a standard pressure as a function of temperature enables numerical values of h to

† See, for example, N. Eumorfopoulos and J. Rai, *Phil. Mag.* **2**, 961 (1926).

be ascribed to each isenthalp. The engineering importance of such data will be readily understood when it is realized how many useful devices, particularly turbines and other heat engines, and gas liquefiers,† depend for their working on the flow of gases. The work of Callendar on the thermodynamic properties of steam‡ may be consulted to illustrate this point. In this connexion it is worth remarking that although we set out in deriving (6·38) to express the Joule-Kelvin effect in terms of what we have called measurable properties, in fact this equation has proved far more useful in the reverse sense. Since the effect only occurs as a result of departures from the perfect gas law it may be, and has been, used to derive better forms of the equation of state for gases whose equation is not easily determined directly, such as steam. Similarly, (6·38) may be employed to estimate corrections to the perfect gas law for nearly perfect gases, and hence to correct gas thermometers to the absolute scale of temperature. We shall discuss this application more fully later in the chapter.

Radiation

As another example of the application of thermodynamics to simple systems we shall consider the properties of the electromagnetic radiation which is present inside a cavity bounded by opaque walls maintained at a uniform temperature (*cavity radiation*). It is not at all obvious that we are justified in applying thermodynamics to such a problem, since the formulation of the laws of thermodynamics was based on experience of material bodies, and we need feel no *a priori* confidence that they are of sufficiently wide validity to embrace mechanical processes in which radiation provides the motive force. The success of thermodynamics in these circumstances is perhaps the strongest evidence we possess for regarding the laws as valid in all physical situations to which they can be applied.

By means of electromagnetic radiation energy may be transported from one place to another; it is on this fact that the application of thermodynamics rests. Anything in the nature of a general thermodynamic theory of radiation is rendered difficult by virtue of the fact that the term *equilibrium* is not normally applicable. The light emitted by a glowing filament is no more in equilibrium than is a stream of molecules escaping from a gas-filled vessel into a vacuum. This is the reason for the peculiar importance of cavity radiation, since it represents an equilibrium situation. Moreover, as Kirchhoff showed, the properties of cavity radiation are such as to make it, if anything, an even simpler thermodynamic system than a fluid. Kirchhoff used the

† For a detailed account of gas liquefaction consult M. and B. Ruhemann, *Low Temperature Physics* (Cambridge, 1937).

‡ H. L. Callendar, *Properties of Steam* (Arnold, 1920).

second law to prove that, provided the walls of the cavity are opaque, the quality of the radiation in equilibrium inside is independent of the nature of the walls and depends only on their temperature. The proof is very simple. Imagine two cavities, A and B, maintained at the same temperature T and connected by a narrow tube through which radiation may pass from one to the other. If the energy transported from A to B exceeds that transported from B to A, it would be possible to raise the temperature of B to $T+\Delta T$ and for B still to receive more energy than it emitted; there would then be a steady flow of energy from a colder to a hotter body and the second law would be violated. It follows, then, that whatever the nature of the walls of A and B the energy flow must be the same from both, and hence the energy density the same in both. By imagining colour filters or polarizers inserted in the connecting tube it can be proved that the radiation is unpolarized and that the spectral distribution of cavity radiation is independent of the walls; it is also easily seen that the radiation must be isotropic, i.e. at no point within the cavity is there any preferred direction either of polarization or of energy flux.†

Thus the quality of cavity radiation depends on the temperature alone, and in particular all the extensive thermodynamic variables of the radiation in a cavity are proportional to the volume, i.e. $U=uV$, $S=sV$, etc., where u, s, etc., are functions of temperature only.

We may now treat the cavity containing the radiation exactly as if it were a vessel containing a gas, and apply the fundamental equation to the system. For this purpose we must know what pressure is exerted on the walls of the cavity by the radiation, and this may be derived by electromagnetic theory.‡ A useful alternative trick which avoids detailed calculations is to employ the photon theory of radiation in conjunction with the elementary kinetic theory of gases. If we think

† The assumption is inherent in this proof that a hole can be cut in the cavity wall which is large enough to allow radiation to escape freely, but not so large as to disturb the radiation in the cavity appreciably. This assumption is no longer valid if the dimensions of the cavity become comparable with the wavelength of the radiation considered. For example at a temperature of 1° K. the energy of cavity radiation is largely concentrated in wavelengths around 1 cm., and in order to allow such energy to escape freely the hole cut must be about 1 cm. or more in diameter. Such a hole would interfere considerably with the radiation in a cavity of only a few cubic centimetres, so that the conditions postulated in Kirchhoff's proof cannot be met. The quality of the radiation in such a cavity is indeed different from that in a very large cavity at the same temperature, and the Stefan-Boltzmann law (6·48) does not apply, nor does Planck's law in the form (6·50). A case of this sort needs special treatment, which, however, is not difficult by the statistical methods used to derive Planck's law.

‡ G. P. Harnwell, *Principles of Electricity and Electromagnetism* (McGraw-Hill, 1938), p. 537.

of the radiation as a gas whose particles all move with the velocity of light, c, and have different energies according to the spectral distribution in the radiation, we may write for the pressure

$$P = \tfrac{1}{3}\rho c^2,$$

in which ρ is the density (mass per unit volume) of the photons. But according to Einstein's mass-energy relation, ρc^2 is simply the density of energy in the gas, so that

$$P = \tfrac{1}{3}u, \tag{6·45}$$

which is the required answer, and the same as derived by electromagnetic arguments. It follows then that the pressure of cavity radiation is also a function of temperature only. The fundamental equation,

$$\mathrm{d}U = T\,\mathrm{d}S - P\,\mathrm{d}V,$$

may now be rewritten, by putting $U = uV$, $S = sV$, $P = \tfrac{1}{3}u$, in the form

$$\mathrm{d}u = T\,\mathrm{d}s + \frac{1}{V}(Ts - \tfrac{4}{3}u)\,\mathrm{d}V.$$

But u and s are functions of temperature only, whence

$$\mathrm{d}u = T\,\mathrm{d}s \tag{6·46}$$

and

$$Ts = \tfrac{4}{3}u. \tag{6·47}$$

From these equations we see that $\mathrm{d}u/\mathrm{d}s = \tfrac{4}{3}u/s$, so that $u \propto s^{\frac{4}{3}}$; hence, from (6·47),

$$u = aT^4, \quad \text{where } a \text{ is a constant,} \tag{6·48}$$

and

$$s = \tfrac{4}{3}aT^3. \tag{6·49}$$

The result expressed in (6·48) is the Stefan-Boltzmann fourth-power law for radiation, which is completely confirmed by experiment.†

To obtain more information about the nature of cavity radiation, such as the spectral distribution of energy, is not possible by pure thermodynamic means. An interesting extension of the thermodynamic result was achieved by Wien, who considered the modifications suffered by the radiation during an adiabatic expansion,‡ but the complete solution of the problem requires the methods of quantum statistics, and leads to Planck's celebrated formula for the

† A more concise proof of this result may be obtained by use of the Gibbs-Helmholtz equation (5·14). Since $\mathrm{d}F = -S\,\mathrm{d}T - P\,\mathrm{d}V$, we have that $(\partial F/\partial V)_T = -P$. But the expansion of radiation at constant temperature only creates new radiation of the same quality, so that $(\partial F/\partial V)_T = f$, the free energy per unit volume. Hence $f = -P = -\tfrac{1}{3}u$; substituting this in (5·14) we find that $4u = T\,\mathrm{d}u/\mathrm{d}T$, or $u \propto T^4$.

‡ G. P. Harnwell and J. J. Livingood, *Experimental Atomic Physics* (McGraw-Hill, 1933), p. 58.

energy density of radiation, $u_\nu \mathrm{d}\nu$, between the frequencies ν and $\nu + \mathrm{d}\nu$,

$$u_\nu \mathrm{d}\nu = \frac{8\pi h\nu^3 \mathrm{d}\nu / c^3}{\mathrm{e}^{h\nu/kT} - 1}, \tag{6·50}$$

which we introduce here solely in order to facilitate the following discussion. It will be seen by integrating over all frequencies that this formula is in accordance with the Stefan-Boltzmann law.

So far we have considered only the radiation in a cavity; we now extend the argument to the radiation emitted by a heated surface. First let us consider an opaque black surface, one which completely absorbs all radiation falling on it from any angle. We have seen that according to Kirchhoff's law a cavity at uniform temperature lined

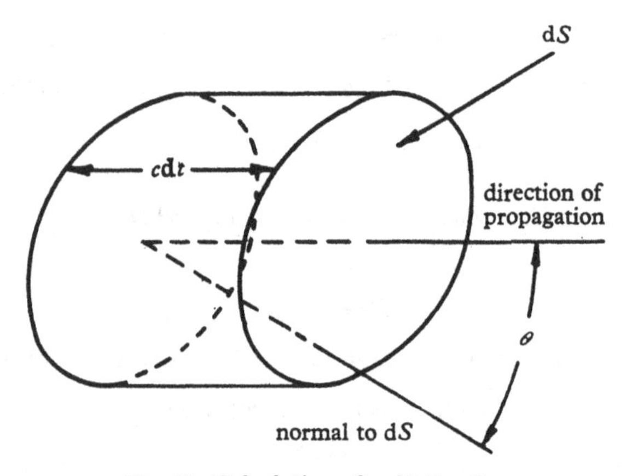

Fig. 19. Calculation of radiation flux.

with such a surface would contain cavity radiation having the spectral distribution given to Planck's formula (6·50). Such a situation can only occur if the surface is emitting spontaneously the same radiation as falls on it,† which may readily be calculated in terms of u_ν, the energy density. Consider an element of the surface $\mathrm{d}S$ (fig. 19); in time $\mathrm{d}t$ the radiation propagated within a solid angle $\mathrm{d}\omega$ which will fall on $\mathrm{d}S$ from angles around θ to the normal will be that contained in a cylinder of side $c\,\mathrm{d}t$ and volume $c\,\mathrm{d}t\,\mathrm{d}S\cos\theta$. The density of radiation in the solid angle $\mathrm{d}\omega$, and frequency range $\mathrm{d}\nu$, is $u_\nu \mathrm{d}\nu\,\mathrm{d}\omega/(4\pi)$, since the

† Here and in what follows we take for granted the truth of Prévost's theory of exchanges, according to which the processes of absorption and scattering on the one hand, and spontaneous radiation on the other, are independent. Such an assumption implies a mechanism not immediately deducible from the observations, so that arguments based on it are, strictly speaking, outside the realm of classical thermodynamics.

radiation is isotropic, so that the radiation falling on the surface is $cu_\nu \cos\theta\, d\nu\, d\omega\, dt\, dS/(4\pi)$, or $cu_\nu \cos\theta\, d\nu/(4\pi)$ per unit solid angle, per unit area of surface, per unit time. We conclude therefore that the radiation emitted by unit area of a black body per unit time, per unit solid angle, is given by the formula

$$E_{\nu,\omega}\, d\nu = \frac{c}{4\pi} u_\nu \cos\theta\, d\nu. \tag{6·51}$$

By integrating over all angles between 0 and $\frac{1}{2}\pi$, we arrive at an expression for the total radiation of the surface in the frequency range $d\nu$,

$$E_\nu\, d\nu = \tfrac{1}{4} c u_\nu\, d\nu, \tag{6·52}$$

and, by integrating over all frequencies,

$$\begin{aligned} E &= \tfrac{1}{4} cu \\ &= \tfrac{1}{4} caT^4 \quad \text{from (6·48).} \end{aligned} \tag{6·53}$$

If the surface is not black, but absorbs only a portion α of the incident radiation, then the emission will be reduced by a factor α from the formulae (6·52) and (6·53) derived for a black body. The emitted radiation will not necessarily have an angular distribution as $\cos\theta$, which necessarily holds for a black body; what is emitted must have such a distribution as will make up for the absorbed radiation and combine with the unabsorbed radiation to produce an isotropic distribution of radiation leaving the surface. For example, a rough surface scatters such radiation as it does not absorb in all directions, more or less with a $\cos\theta$ angular variation, since the projected area of the surface varies as $\cos\theta$; the radiation emitted from the surface will then have approximately a $\cos\theta$ variation also. But it is possible to produce smooth reflecting surfaces which absorb strongly in certain directions only; these will then emit strongly only in those directions. Many different situations are possible with partially absorbing surfaces, but the general rule holds good, that they emit in such a way as to preserve the normal distribution of radiation when they form part of the wall of a cavity. By application of this rule it is possible in principle to determine exactly what will be emitted by any given surface at any temperature if its absorbing and scattering properties are known.

The same method of reasoning may be applied to surfaces which are not perfectly opaque, and we shall illustrate it by considering a layer which does not reflect or refract radiation but which attenuates it so that a fraction f is absorbed, the rest being transmitted. If we place this layer, at temperature T, in front of a black surface at the same temperature, the two together form a black surface at uniform temperature, since all radiation incident is absorbed either by the layer or by the black surface. The composite body thus emits in accordance with (6·51), as does the black surface. But of the radiation emitted by the black surface only a fraction $(1-f)$ penetrates the absorbing layer; it follows that the radiation by the absorbing layer is f times that of a black body. A transparent body is thus a poor radiator, a fact which is well illustrated in laboratory practice when fused silica is worked in a gas flame. The softening temperature of silica is very much higher than can usually be reached with the gas flames used, but silica fortunately has the property of maintaining its transparency over a wide band of optical and infra-red wavelengths up to a high temperature. It is thus heated by molecular bombardment in the flame and has no adequate radiative means of losing its heat. In consequence its temperature rises much higher than that of glass in the same flame. One has only to contaminate the surface of the silica with a little carbon, which is a good radiator, to make it quite impossible for the required working temperature to be reached.

The sun provides a more interesting illustration of this same principle. If, instead of its visible radiation, the radiation which it emits at radio-frequencies corresponding to wavelengths about 5 m. is examined, the sun appears to be several times larger than its visible diameter. Moreover, the intensity of radiation suggests that the emitter is a black body with a temperature as high as one million degrees, enormously greater than the temperature of the visible surface, about 6000° C. If shorter radio waves are studied in the same way there is observed a marked fall in apparent size and temperature of the sun. These observations are explicable by the hypothesis, to which other phenomena give strong support, that the visible sphere of the sun is surrounded to a distance of many times its diameter by a highly rarefied atmosphere, consisting mainly of protons and electrons, at an extremely high temperature. Such an ionized gas has an absorbing power for radiation which varies as the square of the wavelength. It may therefore be almost completely transparent to visible light, and opaque and 'black' to radio waves. Correspondingly it may be a powerful emitter of radio waves and yet so weak an optical emitter as only to be visible, as the corona, at a time of total solar eclipse.

Finally, before leaving the topic of radiation, it is of interest to note an application of the foregoing results at such low frequencies that transmission lines and lumped circuit components become practicable. We may imagine two large cavities, maintained at a uniform temperature T, which are connected by means of radio aerials terminating a transmission line, as shown schematically in fig. 20. The aerials, which may take any form or size, are supposed to be matched to the transmission line, i.e. if power is sent along the line towards one of the aerials, it is all radiated and none is reflected. Then Kirchhoff's argument may be repeated to show that the power picked up by each aerial, and hence by any matched aerial, is the same. If the aerial is imperfectly matched, and radiates only a fraction f of the power sent along the line, reflecting the rest, then it is easily seen that its capacity to pick up power from the radiation field in the cavity is also reduced by a factor f. By considering in detail a suitably simple aerial it is possible to show that when matched to its line it will pick up in the range of

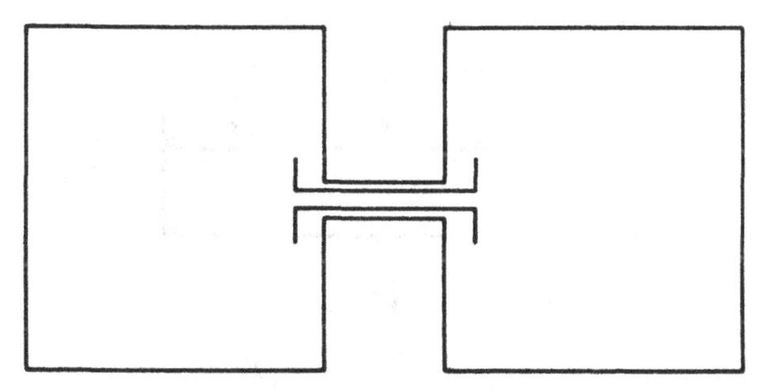

Fig. 20. Temperature enclosures connected by transmission line.

frequency $\mathrm{d}\nu$ power $W\,\mathrm{d}\nu$ equal to $c^3 u_\nu \mathrm{d}\nu/(8\pi\nu^2)$, which from (6·50) is seen to be given by the expression

$$W\,\mathrm{d}\nu = \frac{h\nu\,\mathrm{d}\nu}{e^{h\nu/kT}-1}$$

$$\approx kT\,\mathrm{d}\nu \quad \text{when} \quad h\nu \ll kT. \tag{6·54}$$

The approximation (6·54), which is normally valid for radio-frequency phenomena, is equivalent to the use of the Rayleigh-Jeans radiation formula ($u_\nu \mathrm{d}\nu = 8\pi kT\nu^2 \mathrm{d}\nu/c^3$) rather than that of Planck.

Now let us replace one of the aerials in fig. 20 by a resistor which matches the transmission line and which is maintained at temperature T. Then the theorem of Kirchhoff shows that the resistor must generate power at the rate given by (6·54). This result is the celebrated Nyquist formula for the 'Johnson noise' spontaneously generated by a resistor. It will be seen that this noise is the exact analogue of the radiation emitted by a black body. But what is more important, and the main reason for our introducing the topic, we may see from this example how a typical 'fluctuation' phenomenon, Johnson noise, which may be thought of as originating in the Brownian motion of the electrical carriers, may be related to an 'equilibrium' phenomenon, the radiation in equilibrium in a cavity. This example shows clearly, what we have stressed before, that fluctuations are not to be regarded as spontaneous departures from the equilibrium configuration of a system, but are manifestations of the dynamic character of thermal equilibrium itself, and quite inseparable from the equilibrium state.

Surface tension and surface energy

The elementary facts of surface tension are well known and we shall base the following discussion on what is probably the clearest-cut fact

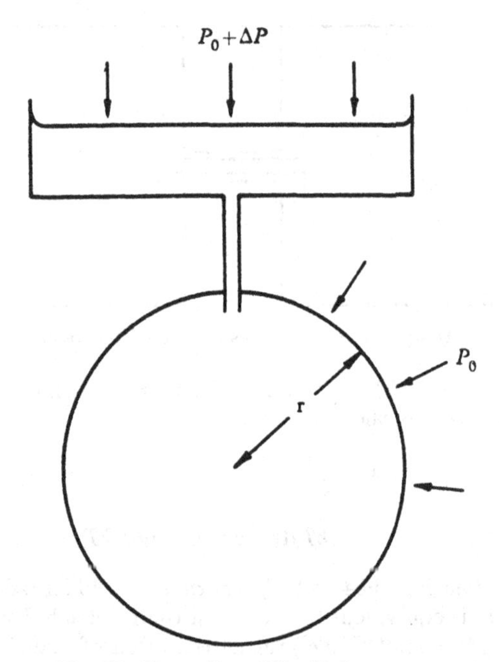

Fig. 21. Formation of liquid drop.

of all, that the pressure inside a spherical surface of a liquid is greater than that outside by an amount $2\sigma/r$, where σ is the surface tension and r the radius of the surface. We shall concern ourselves only with the surface behaviour of liquids and say nothing of the properties of soap films.

Let us now consider a hypothetical experiment, illustrated in fig. 21, in which a spherical drop is formed from the bulk liquid at the end of a very narrow pipette. We assume the liquid to be incompressible, so that its free energy per unit volume is independent of pressure (since $(\partial F/\partial P)_T = -P(\partial V/\partial P)_T = 0$). The reader may readily extend the argument to apply to a compressible liquid. If the atmospheric pressure is P_0, the pressure required on the plane surface of the reservoir is $P_0 + \Delta P$, where $\Delta P = 2\sigma/r$. Let us now increase the radius of the drop reversibly by an amount $\mathrm{d}r$, keeping the temperature constant. A volume $4\pi r^2 \mathrm{d}r$ of liquid must be forced through the pipette, and the work required is $4\pi r^2 \mathrm{d}r \Delta P$, i.e. $8\pi r\sigma \mathrm{d}r$. This is equal (see p. 56) to the increase of free energy of the system. But the free energy of the reservoir has decreased by an amount $4\pi r^2 f_0 \mathrm{d}r$, where f_0 is the free energy per unit volume of the bulk liquid. Therefore

$$\left(\frac{\partial F_{\text{drop}}}{\partial r}\right)_T = 4\pi r^2 f_0 + 8\pi r\sigma,$$

i.e.

$$F_{\text{drop}} = \frac{4\pi}{3} r^3 f_0 + 4\pi r^2 \sigma$$

$$= V f_0 + \mathscr{A}\sigma, \tag{6·55}$$

where V is the volume of the drop and $\mathscr{A}$ its surface area. The free energy of the drop is thus made up of two terms, one proportional to the volume and one to the area, and the surface tension may be interpreted as the free energy per unit area of surface.

The molecular reason for the appearance of surface tension is made clear by this result. The forces between the molecules of a liquid extend only over a short range, and at a distance within the surface greater than this range the molecular behaviour is unaffected by the proximity of the surface, and the free-energy density of an element of volume is in consequence independent of the position of that element. Thus in a mass of liquid whose dimensions are much greater than the range of molecular forces the major contribution to the free energy is a term proportional to the volume, the first term of (6·55). Close to the surface, however, the molecular behaviour is affected by the surface, and the second term of (6·55) represents the correction needed for this effect. From this argument it will be seen that the surface contribution to the free energy is not solely a property of the liquid state of matter, but may be expected to exhibit itself in solids and even gases, though only in liquids does it give rise to the straightforward observable behaviour characteristic of surface tension. As will be shown in Chapter 7 (p. 107), the equilibrium state of a system maintained at constant volume and temperature is that of minimum free energy, and the tendency for a liquid drop to assume a spherical shape is a simple manifestation of this rule, since a sphere has the minimum surface area for a given volume. In a crystalline solid, the surface contribution to the free energy may vary widely for different orientations of the surface with respect to the crystalline axes, and the equilibrium shape of a small crystal is not now spherical, but of such a form as exposes most fully the crystal faces of smallest free energy. This accounts in general terms for the well-marked, and often complicated, shapes into which solids crystallize naturally from the vapour or from solution.†

So long as the dimensions of the body considered are much larger than the range of molecular forces, and provided there are no surfaces whose radii of curvature are smaller than this range, the result (6·55) may be expected to be valid. But it cannot be applied when two surfaces are so close together that there is direct molecular interaction between them; the distinction between volume and surface contributions becomes inapplicable. Similarly, at a sharp edge it may be expected that additional corrections to the free energy will be needed, by addition of a term proportional to the length of the edge, and there may be yet another term to be applied where three or more surfaces meet in a point. It must be remembered, however, that these terms will depend on the angles between the surfaces. There are few phenomena in which they play any significant part and we shall consider them no further.

The surface contribution to the free energy finds of course its analogues with the other thermodynamic functions, and there will be, in general, surface contributions to the internal energy, entropy, etc. It is easily seen that these quantities are related by expressions analogous to those relating the thermodynamic functions of complete systems, e.g. if λ is the internal energy per unit area of surface and η the entropy per unit area,

$$\eta = -\mathrm{d}\sigma/\mathrm{d}T \tag{6·56}$$

and

$$\lambda = \sigma + T\eta = \sigma - T\frac{\mathrm{d}\sigma}{\mathrm{d}T}, \tag{6·57}$$

which is the analogue of the Gibbs-Helmholtz equation (5·14). It is of interest to note the consequences of Eötvös's rule when it is applied to these results. Over a very wide temperature range in a liquid, according to this empirical rule, the surface tension of a liquid in contact with its vapour is proportional to $T_c - T$, where T_c is the critical temperature. This implies the rather remarkable fact that η and λ are independent of temperature. The rule cannot, however, be valid either near T_c or at very low temperatures. For at the critical point the liquid and vapour phases become indistinguishable, so that the phase boundary ceases to exist and there can be no entropy contribution from it. We therefore

† This is a considerable over-simplification of the real state of affairs, for a crystal is normally not in its state of minimum free energy, and its shape may be determined more by the detailed processes of its growth than by thermodynamic conditions. See F. C. Frank, *Advanc. Phys.* **1**, 91 (1952).

expect that the surface tension eventually falls to zero with a vanishing gradient, and this seems to be confirmed by experiment; at any rate, Ramsey and Shields have suggested that the critical temperature in Eötvös's rule should be replaced by a temperature about 6° lower for a great many liquids to improve the agreement with experiment. A direct measurement of σ within a few degrees of the critical temperature is very hard to carry out. If the liquid phase can persist to temperatures approaching the absolute zero (as with liquid helium) we may expect that σ will tend to become independent of temperature, since according to the third law the surface contribution to the entropy must vanish at 0° K. This behaviour is indeed shown by liquid helium.

The surface contributions to the thermodynamic functions differ from the functions of the majority of complete systems in that it is possible to ascribe absolute values to them without the normal ambiguity of additive constants. This is because the surface area may be varied at will without alteration of the volume. In principle therefore one may determine experimentally the work or heat required to increase the area of surface by a known amount and thus determine unequivocally the free energy, entropy, etc., per unit area. It will be noted that the same property is possessed by cavity radiation.

Establishment of the absolute scale of temperature

The establishment of the absolute scale of temperature involves, in principle, the calibration of a convenient empirical thermometer in terms of absolute temperature, and the systems which we have analysed in this chapter provide four different methods of doing this over limited ranges of temperature. We shall not concern ourselves here with any technical details, which are often of considerable complexity on account of the great refinement needed to ensure high accuracy. The details may be found by reference to standard works concerned with the problems of thermometry.† It is, however, worth mentioning that the aim of establishing the absolute scale is not simply to possess one empirical thermometer and a calibration curve for it so that the owner may be able to measure temperature absolutely. It is also desirable that replicas may be constructed in other laboratories and calibrated without a complete repetition of the arduous experiments involved in calibrating the first standard thermometer. For this reason much work has gone into finding suitable materials for substandard thermometers, whose properties shall be sufficiently reproducible and sufficiently smooth functions of temperature that calibration at a few fixed points will be adequate to enable the complete calibration curve to be inferred. An example of such a

† See references on p. 48.

thermometer is the platinum resistance thermometer. No two specimens of platinum have resistances which vary in exactly the same way with temperature, but it is known from comparison of many specimens that allowance may be made for variation by calibration at a few quite widely spaced fixed points, since the resistance over a wide range is very nearly a linear function of absolute temperature. Once suitable thermometric substances have been developed there remains the task of discovering accurately reproducible fixed points and determining their absolute temperatures as well as possible. The melting-point of ice (273·15° K.) is fixed by definition, and the boiling-point of water (373·15° K.) by accurate measurement; other melting-points (or, preferably, triple points) and boiling-points have been added to extend the range of temperature both upwards (b.p. of sulphur, 717·8° K.; m.p. of gold, 1336° K.) and downwards (sublimation point of CO_2, 194·7° K.; b.p. of oxygen, 90·2° K.; triple point of hydrogen 14·0° K.).

We shall discuss briefly the thermodynamic principles involved in four methods:

(1) correction of the gas scale by extrapolation to zero pressure;
(2) correction of the gas scale by means of the Joule-Kelvin effect;
(3) radiation pyrometry;
(4) calibration of the magnetic scale below 1° K.

Of these we shall dismiss the third in a few words. The Stefan-Boltzmann fourth-power law enables the temperature within a cavity to be determined from the total radiation emitted from a small hole cut in the cavity wall. The thermodynamic principles have already been discussed sufficiently fully; the technical details, and formidable experimental difficulties which must be overcome, will not be entered into. No great accuracy (better than about 1°) has been attained in radiation pyrometry, but it is the only method of temperature measurement which can be carried into the region of very high temperatures (over 2000° K.) and still pretend to be related to the absolute scale.

The gas thermometer, on the other hand, is capable of very high accuracy (about 0·01° or better) in the measurement of empirical temperature, so that the methods of making the small corrections to reduce its empirical scale to the absolute scale are of great importance in thermometry. As we saw in Chapter 5, a perfect gas, obeying Boyle's and Joule's laws, needs no calibration, since, if used in a constant-volume gas thermometer, the pressure is exactly proportional to the absolute temperature. No real gas is perfect, of course, and we shall now extend the analysis of Chapter 5 to show that the extrapolation to zero pressure of the properties of a real gas is indeed justified, in that it yields a measure of the absolute temperature. For this purpose we assume, as is found experimentally, that the departures of a real gas from Boyle's and Joule's laws may be expressed as power series in the pressure

$$PV = A + BP + CP^2 + \dots \tag{6·58}$$

and

$$U = \alpha + \beta P + \gamma P^2 + \dots, \tag{6·59}$$

in which the coefficients A, B, α, β, etc., are functions of temperature only. Then from the fundamental equation and M.3 we may write

$$\left(\frac{\partial U}{\partial P}\right)_T = -T\left(\frac{\partial V}{\partial T}\right)_P - P\left(\frac{\partial V}{\partial P}\right)_T,$$

or, by use of (6·58) and (6·59),

$$\beta + 2\gamma P + \dots = \left(A - T\frac{\mathrm{d}A}{\mathrm{d}T}\right)\frac{1}{P} - T\frac{\mathrm{d}B}{\mathrm{d}T} - \left(C + T\frac{\mathrm{d}C}{\mathrm{d}T}\right)P - \dots.$$

Hence, comparing coefficients, we see that

$$A - T\mathrm{d}A/\mathrm{d}T = 0,$$

or

$$A \propto T.$$

Thus the value of PV does indeed tend, as P tends to zero, to a value which is proportional to T, and the extrapolation of a real gas to zero pressure yields correct values of the absolute temperature.

The process of extrapolation, however, involves in practice precise measurements with a gas thermometer at a number of different pressures, and this procedure may be largely avoided by use of subsidiary determinations of the Joule-Kelvin coefficient of the thermometric gas. It will be recalled that the Joule-Kelvin effect vanishes for a perfect gas, so that it gives directly a measure of the imperfection of the gas. It is not in itself sufficient, as will be understood by recalling the inversion phenomenon; the absence of a Joule-Kelvin effect does not guarantee the perfection of the gas. We must therefore look more closely into the matter to see what use can be made of the Joule-Kelvin effect and what additional information is needed. According to (6·38), the Joule-Kelvin coefficient $(\partial T/\partial P)_h$ (which we shall denote by μ), is expressible in the form,

$$\mu c_p = T(\partial v/\partial T)_p - v. \tag{6·60}$$

Here all temperatures are absolute, but in any experiment on the effect they are measured in terms of an empirical scale. Let us define the empirical temperature in terms of a given constant-volume gas thermometer, putting $\theta = aP$, where P is the pressure in the thermometer and a is a constant, which may be chosen so that there are 100°

between the melting-point of ice and the boiling-point of water. Then to express (6·60) in terms of θ rather than T, we note that

$$\mu' \equiv \left(\frac{\partial\theta}{\partial P}\right)_h = \mu\frac{\mathrm{d}\theta}{\mathrm{d}T}, \tag{6·61}$$

$$c'_P \equiv \left(\frac{\partial h}{\partial\theta}\right)_P = c_P\frac{\mathrm{d}T}{\mathrm{d}\theta} \tag{6·62}$$

and

$$\left(\frac{\partial v}{\partial T}\right)_P = -\left(\frac{\partial P}{\partial\theta}\right)_v\left(\frac{\partial v}{\partial P}\right)_\theta\frac{\mathrm{d}\theta}{\mathrm{d}T}. \tag{6·63}$$

If the Joule-Kelvin coefficient is measured with the same gas and under the same conditions as exist in the gas thermometer, the term $(\partial P/\partial\theta)_v$ in (6·63) is by definition P/θ. Hence, combining equations (6·60)–(6·63), we have

$$\mu' c'_P = -v\left\{\frac{PT}{v\theta}\left(\frac{\partial v}{\partial P}\right)_\theta\frac{\mathrm{d}\theta}{\mathrm{d}T}+1\right\}.$$

If c'_P is the specific heat per unit mass, so that $v = 1/\rho$, ρ being the density of the gas,

$$T\frac{\mathrm{d}\theta}{\mathrm{d}T} = -\theta(1+\rho\mu' c'_P)\left(\frac{\partial(\log P)}{\partial(\log v)}\right)_\theta = \mathscr{F}(\theta). \tag{6·64}$$

The function on the right-hand side may be regarded as a function of θ only, since all the quantities entering therein are prescribed to be measured under the conditions which prevail in the gas thermometer. It will be observed that everything that occurs in $\mathscr{F}(\theta)$ is a measurable quantity, and that the information needed besides the value of μ' consists of the density and specific heat of the gas, and the quantity $-\left(\frac{\partial(\log P)}{\partial(\log v)}\right)_\theta$ which is the isothermal bulk modulus of the gas divided by its pressure. If the gas is perfect the last is equal to unity and μ' vanishes, so that $\mathscr{F}(\theta) = \theta$ and $\theta = T$. If the gas is not quite perfect, as is the practical situation, $\rho\mu' c'_P$ is much smaller than unity, so that the precision of measurement required for these quantities is not so high as the precision desired in correcting the gas scale. On the other hand, the bulk modulus must be measured very accurately, but this is fortunately exactly what can be done in a well-designed gas thermometer.

Equation (6·64) may be integrated in the form

$$\log\left(\frac{T_1}{T_0}\right) = \int_{\theta_0}^{\theta_1}\frac{\mathrm{d}\theta}{\mathscr{F}(\theta)}. \tag{6·65}$$

If $\mathscr{F}(\theta)$ is measured between 0 and 100° C. this integration may be performed numerically to give the ratio of the absolute temperatures at the boiling-point and melting-point, and thus to fix the value of the absolute zero on the Celsius scale (or to determine the melting-point of ice on the absolute Centigrade scale). The relation between any other T and θ is then found by integrating from the melting-point to θ, and hence the gas scale is calibrated in terms of the absolute scale.

Lastly we turn to the problem of establishing the temperature scale in the range of temperatures below 1° K. which are reached by adiabatic demagnetization of paramagnetic salts. Gas thermometry is impossible in this range since even helium has so low a vapour pressure (e.g. 10^{-32} cm. of mercury at 0·1° K.) that no useful measurements on the gaseous state can be made. The helium gas thermometer has been used to establish the absolute scale with an accuracy of about 0·01° or better down to 2° K., and the scale has been extended to 1° K. by means of paramagnetic salts which are known, from their behaviour in demagnetization experiments, to obey Curie's law accurately down to 1° K., so that the reciprocal of the susceptibility may be taken as a measure of T. In this range it is convenient to use as an empirical thermometer the vapour pressure of helium, and it is this which is determined as a function of T by means of a gas thermometer or a paramagnetic salt. We shall therefore assume that any temperature of 1° K. or higher whose value is required in establishing the absolute scale below 1° K. may be determined absolutely by reference to a table of helium vapour pressures. We shall not expect to achieve greater relative accuracy below 1° K. than has been achieved above 1° K., since the calibrations are interdependent.

As an empirical thermometer below 1° K. the susceptibility of a paramagnetic salt serves admirably, and the salts which have been used are those which obey Curie's law down to very low temperatures and which, by virtue of this fact, attain a very low temperature in a demagnetization experiment. The empirical ('Curie') temperature, which is customarily denoted by T^*, is defined by the equation

$$T^* \equiv \text{constant}/\chi, \tag{6·66}$$

the constant being chosen so that T^* tends to equal T as the temperature is raised to 1° K. or over. The problem now is to determine the relation between T^* and T at lower temperatures. A number of different methods have been devised, of which we shall discuss only one, which has been used successfully to calibrate several different salts. This method starts from the fact that (see (6·21)), in the absence of a magnetic field,

$$\begin{aligned} T &= \mathrm{d}U/\mathrm{d}S \\ &= \left(\frac{\mathrm{d}U}{\mathrm{d}T^*}\right) \bigg/ \left(\frac{\mathrm{d}S}{\mathrm{d}T^*}\right). \end{aligned} \tag{6·67}$$

The experiments carried out are aimed at determining the two differential coefficients of (6·67) independently over a range of Curie temperatures.

The coefficient (dU/dT^*) is clearly the specific heat of the salt in zero magnetic field measured on the Curie scale, and is determinable in principle by demagnetizing the salt to a low temperature and measuring the change of susceptibility when a known amount of heat is supplied. The chief difficulty in practice is to ascertain the rate of supply of heat which, it may be mentioned, should be supplied uniformly to the salt on account of the extremely poor thermal conductivity of the salt at these low temperatures. In the earliest work the heat was provided by mixing a little γ-radioactive material with the salt, and determining the amount of γ-ray energy absorbed in a separate experiment at higher temperatures. More recently an alternative means of heating has been devised which makes use of the hysteresis losses which the salt suffers when subjected to an alternating magnetic field; any heat production in the salt will show up as a resistive component in the inductance of the coil which is used to measure χ, and the heat supplied may be determined from the magnitude of this resistance and the current in the coil. There are other difficulties in determining (dU/dT^*), mainly concerned with unavoidable heat leaks to the salt and desorption of helium gas from its surface on warming, but these only limit the accuracy attainable without invalidating the method, and we shall discuss them no further.

A quite different procedure is used to measure (dS/dT^*) as a function of T^*. The first stage involves the determination of S as a function of $\mathscr{H}$ at some convenient temperature T_0 in the range above 1° K., where the absolute value of T_0 may be found. This is carried out by making use of the analogue of M.3,

$$\left(\frac{\partial S}{\partial \mathscr{H}}\right)_T = \left(\frac{\partial \mathscr{M}}{\partial T}\right)_{\mathscr{H}}.$$

By studying the magnetization of the salt as a function of $\mathscr{H}$ over a range of temperatures around T_0 the differential coefficient on the right may be determined as a function of $\mathscr{H}$ at temperature T_0, and hence by numerical integration the curve of S against $\mathscr{H}$ is plotted. The second stage consists of a series of adiabatic demagnetizations of the salt, starting always from temperature T_0 but varying the starting value of the field. From the first set of measurements the starting value of the entropy is known, and this is also (ideally) the final value, since the demagnetization is adiabatic. It is therefore possible, by measuring the value of T^* when the field reaches zero, to find how S varies with T^* and hence to construct a curve of the variation of (dS/dT^*) with T^*. From measurements such as these a curve showing the relation between T^* and T may be constructed, of which fig. 22 is an example, the salt here being chrome alum. The inaccuracies of the measurements are mainly due to imperfect realization of adiabatic conditions, but it will also be appreciated that there is a considerable amount of

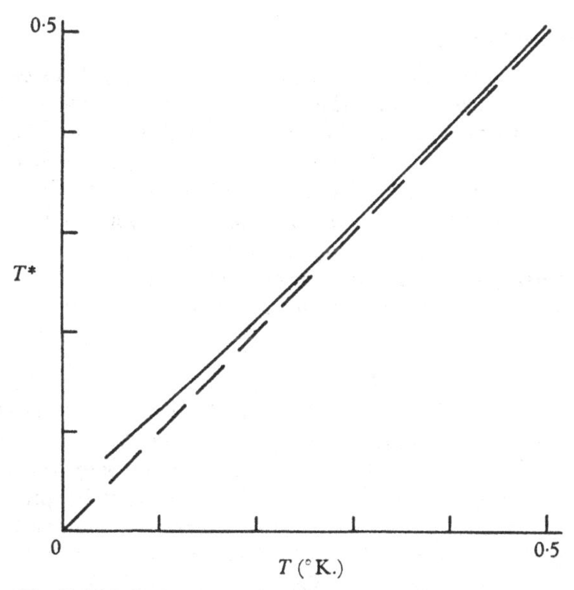

Fig. 22. T-T^* relation for chrome alum (E. Ambler and R. P. Hudson, *Rep. Progr. Phys.* **18**, 251, 1955).

numerical manipulation of the data and smoothing out of experimental scatter involved, and that the errors in these procedures are liable to be cumulative. It is thus not surprising that very high accuracy has not been attained, and that different workers do not always agree about the detailed shape of the curve, particularly at the lower end of the temperature range. Nevertheless, the measure of consistency achieved at such an extreme of temperature is quite a triumph of applied thermodynamics.

CHAPTER 7

THE THERMODYNAMIC INEQUALITIES

The increase of entropy

We have so far discussed the consequences of the second law only in so far as they are concerned with the existence of entropy and absolute temperature. In this chapter we shall begin the discussion of Clausius's inequality (4·14), which holds the clue to the difference between reversible and irreversible processes and provides a criterion by which it may be determined whether a given process is physically possible.

According to Clausius's inequality, for any closed cycle of a system, during which heat may be supplied or extracted by an external body at temperature T (not necessarily constant during the change),

$$\oint q/T \leqslant 0.$$

The equality sign expresses the situation which obtains when the cycle takes place in a reversible manner. If in any change the equality sign is found to hold, there is no thermodynamic reason why the cycle should not be exactly reversed, although there may be practical difficulties in the way of accomplishing the reversal, and we shall refer to all such cycles as reversible without considering the matter of practicability. Let us consider a cycle which may be decomposed into two parts, the first an irreversible change from one equilibrium state A of the system to another equilibrium state B, and the second a reversible return to the state A. In this cycle $\oint q/T < 0$, so that

$$\int_{A \text{ irrev.}}^{B} q/T + \int_{B \text{ rev.}}^{A} q/T < 0.$$

Now for the reversible change represented by the second integral we may replace $\int_{B \text{ rev.}}^{A} q/T$ by $S_A - S_B$, since this is how an entropy change is defined. Hence

$$\int_{A \text{ irrev.}}^{B} q/T < S_B - S_A. \tag{7·1}$$

We may simplify the notation and include reversible changes by writing (7·1) in the form

$$\int q/T \leqslant \Delta S, \tag{7·2}$$

or, for a differential change,

$$T\,\mathrm{d}S \geqslant q. \tag{7·3}$$

The inequality (7·2) is the fundamental thermodynamic inequality; once again we stress that the temperature T which occurs in (7·2) and (7·3) is the temperature of the body which supplies the heat, and not the temperature of the system undergoing the change.

If the system is thermally isolated, so that $q=0$, (7·2) takes a particularly simple form,

$$\Delta S \geqslant 0, \tag{7·4}$$

which expresses the *law of increase of entropy*:

The entropy of an isolated system can never diminish.

This law, which we shall refer to as the entropy law, provides the thermodynamic criterion for deciding which processes that could conceivably occur in an isolated system (without violating the first law) can actually be effected without violating the second law. An examination of simple examples will show the operation of this law. Consider first two bodies at different temperatures, T_1 and T_2 ($T_1 > T_2$), and let them be brought momentarily into thermal contact so that heat q flows from the hotter to the colder; the entropy of the hotter body decreases by q/T_1, while that of the colder increases by a greater amount q/T_2, and the resultant total effect is that the entropy of the system increases. The reverse process, of heat flowing from the colder to the hotter body, is excluded by the law.† Secondly, consider a quantity of gas contained within a vessel which is surrounded by an evacuated space. If the vessel is pierced the gas expands irreversibly to fill the whole space, and its entropy is thereby increased (see p. 69). As a third example consider a moving body coming to rest under the influence of frictional forces. The decrease of kinetic energy is accompanied by an increase in temperature, and consequently in entropy, of the body and its surroundings.

† It will be seen that the second law provides a proof of what we have hitherto regarded as a consequence of the converse of the zeroth law, that it is possible to construct a temperature scale (of which the absolute scale is an example) such that there is a monotonic correspondence between hotness and temperature. It is in fact possible to develop thermodynamics without introducing the converse of the zeroth law, if the second law is formulated in such a way that the concepts of *hotter* and *colder* do not enter; Kelvin's and Carathéodory's formulations fulfil this condition, but it would be harder to avoid implicit assumption of the converse of the zeroth law if Clausius's formulation were taken as a basis. Although the treatments of the second law given in Chapter 4 avoided the use of Clausius's formulation, it has been felt to be desirable to state the converse to the zeroth law explicitly at an early stage, since the ideas of *hotness* and *coldness* are so much part of our intuitive perception that they should be placed on a more rigorous basis as early as possible, and not left to follow as a trivial consequence of the much more sophisticated second law.

In the first two examples the changes treated involved the transition from one equilibrium state to another, and were effected by altering the constraints imposed upon the systems, in the first by removal of an adiabatic wall and in the second by altering the volume to which the gas was confined. The third example is slightly different, since the initial state was not an equilibrium state, but in this particular case no difficulty arises if we define the entropy of a moving body, which is in equilibrium in a coordinate frame moving with its centre of mass, to be the same as if it were at rest (at any rate so long as the velocity is small in comparison with the velocity of light; relativistic thermodynamics requires special treatment which will not be attempted here). In all the examples, then, the initial and final entropies are well defined, and the operation of the entropy law is clear. It should not be necessary to point out that the entropy law cannot be applied to any transition in which the entropies of the states concerned are not definable, and this means that, except for trivial exceptions like the third example, it is not applicable except to transitions between equilibrium states. Now for any given set of constraints a thermodynamic system has only one true equilibrium state,† and we may therefore formulate the entropy law in a slightly different way:

It is not possible to vary the constraints of an isolated system in such a way as to decrease the entropy.

This formulation focuses attention on the constraints to which a system is subjected, and it is instructive to follow up this line of argument in connexion with the fluctuations which are an essential feature of the equilibrium state. To make the meaning clear we shall consider a specific example, the second of the three mentioned above, in which a gas is released from a smaller into a larger volume. When the gas is in equilibrium in the larger volume its density is very nearly uniform, but is subject to continual minute fluctuations. Very occasionally larger fluctuations will occur, and there is a continuous spectrum of possible fluctuations ranging, with decreasing probability, from the very small to the very large; so that it is a theoretical possibility (though it is overwhelmingly improbable of observation even on a cosmic time scale) that the gas may spontaneously collapse into the smaller volume from which it originally escaped at the piercing of the wall. It will subsequently expand again to fill the full volume at just the same rate as at the first escape. We may now inquire what

† The existence of 'metastable' states of the sort discussed in Chapter 2, e.g. a mixture of hydrogen and oxygen, does not necessarily invalidate the argument, since we have elected to treat these exactly as if they were stable states, that is, to ignore the possibility that a chemical reaction could occur. See, however, the second paragraph of the next page.

happens to the entropy of the gas during this large-scale fluctuation, and to this question the only satisfactory answer is the perhaps surprising one—nothing. For the continuous spectrum of fluctuations of all magnitudes is, as stressed before, part of the nature of thermodynamic equilibrium; the huge fluctuation just envisaged does not represent a departure from equilibrium—it is simply an extremely rare configuration of the gas molecules, but still just one of the enormous number of different configurations through which the gas passes in its state of equilibrium subject to given constraints. If we ascribe a definite value to the entropy of the gas in equilibrium we must ascribe it not to any particular, most probable, set of configurations, but to the totality of configurations of which it is capable. Thus we see that the entropy (and of course other thermodynamic functions) must be regarded as a property of the system and of its constraints, and that once these are fixed the entropy also is fixed. Only in this sense can any meaning be attached to the statement that the entropy of an isolated mass of gas, confined to a given volume, is a function of its internal energy and volume, $S=S(U, V)$. It follows from this that when the gas is confined to the smaller volume it has one value of the entropy, when the wall is pierced it has another value, and that it is the act of piercing the wall and not the subsequent expansion that increases the entropy. In the same way when two bodies at different temperatures are placed in thermal contact by removal of an adiabatic wall, it is the act of removing the wall and not the subsequent flow of heat which increases the entropy. It will be seen then that our second statement of the entropy law has much to recommend it in that it concentrates upon the essential feature of a thermodynamic change, the variation of the constraints to which a system is subjected.

To take this argument to its ultimate logical conclusion leads to a rather curious situation. Since no walls are absolutely impervious to matter or to heat we may consider that no constraints are perfect; no two bodies in the universe are absolutely incapable of interaction with one another. Therefore the entropy of the universe is fixed once and for all, and the present state of the universe either is, or for thermodynamic purposes simulates, an enormous fluctuation from the mean state of more or less uniform density and temperature. But, apart altogether from the entirely unjustifiable assumption that the universe can be treated as a closed thermodynamic system, this point of view is not very useful, since it makes it difficult, if not impossible, to apply the entropy law in any situation. It is better by far to make a reasonable compromise, of the same nature as those which we made in Chapter 2. Although truly adiabatic walls do not exist, we imagine for the sake of argument that they do, so that small portions of the universe may be considered in isolation. A similar compromise was involved in our discussion of metastability, in which we concluded that no harm would arise from assuming that reactions which proceed immeasurably slowly are not proceeding at all. We are then enabled to define the entropy of physically interesting systems, and apply the entropy law to them without difficulty.

The point of view that it is the constraints, rather than the immediate configuration, which determine the entropy is satisfactory in that it enables fluctuations to take their natural place in the thermodynamic scheme, but it carries one slightly unfortunate consequence with it, that the entropy law is no longer universally valid. A typical violation of the law is exemplified by the following experiment. Let two bodies at different temperatures be initially isolated from one another and then be brought into thermal contact; before they have had time to reach the same temperature isolate them again. In the first stage the entropy is increased by removing the adiabatic wall, in the second it is reduced once more, perhaps to nearly its original value, and the second stage can be thought of as a violation of the entropy law, which can be made as great a violation as we please by choosing the initial temperatures to be as far apart as desired. It will be observed, however, that in the complete experiment the entropy is increased, since we cannot increase the temperature difference between the bodies. Thus no useful decrease in entropy is achieved, and no violation of the second law can be effected by means of such a violation of the entropy law. We conclude therefore that our deduction of the entropy law from the second law is not logically flawless.

The reason for this is to be found in an inconsistency in our point of view concerning large-scale fluctuations. In the development of the laws of thermodynamics we took it for granted that no fluctuations would lead to any observable temperature difference between two bodies in equilibrium, so that we were able to say that bodies in equilibrium were characterized by the same temperature. Now, however, we have committed ourselves to the view that the entropy is determined by the constraints, so that bodies in thermal contact but not at the same temperature are treated as if they were indulging in a huge fluctuation from the average configuration, of such a magnitude as never to have the remotest chance of being observed as a spontaneous fluctuation. If we were to be consistent we should have to make a clear distinction between the temperature of the combined system of two bodies and the temperatures of the bodies separately. The former would be, like the entropy, a function of the constraints and invariant in even the largest fluctuations; the latter would be variable, and only rather imperfectly defined, just as the entropy of one of the bodies is not well defined so long as it is in thermal contact with the other. This discussion is, however, leading us into rather deep waters, and it does not seem useful to carry it any further within the framework of classical thermodynamics. It is by no means easy to incorporate fluctuations consistently, and a full treatment is only possible with the aid of a microscopic picture and the techniques of statistical thermodynamics.

The difficulties which we have just been discussing need not force us to the conclusion that we are wrong to regard the entropy as determined by the constraints. Apart from the type of useless violation of the entropy law mentioned above, the law, in both formulations given, is entirely valid in practice, and we need not hesitate to use either formulation as the basis for determining what changes are permitted by the second law. As a matter of fact, the entropy law has a much wider validity than might be supposed from the foregoing discussion, from which it might have appeared likely that fluctuations lay outside the realm of the entropy law. But it seems most probable that spontaneous fluctuations cannot be used in any way to violate either the entropy law or the second law. One of the earliest discussions of this topic was by Maxwell, who hypothesized a 'demon' having the necessary endowments to enable him to violate the second law. Maxwell imagined two vessels of gas at the same temperature separated by an adiabatic wall, in which was a small hole of little more than molecular dimensions. The demon controlled a trapdoor in such a way as to let through the hole from left to right only such molecules as had more than the average velocity and from right to left only such as had less than the average. In this way he was eventually able to raise the temperature of the gas on the right and lower that of the gas on the left, and so decrease the entropy of the gas as a whole. This hypothetical experiment is essentially one which systematically employs fluctuations (in this case fluctuations of the energy of the gas molecules in a given region) to violate the second law. Now it is implicit in Maxwell's discussion that the entropy of the demon need not enter the problem, and recently Brillouin† has pointed out how unjustified this assumption is. On either side of the adiabatic wall the temperature is uniform, and in consequence the radiation in the vessel is isotropic. Therefore the demon cannot distinguish the form or position of any object in the vessel, and cannot tell when to open or close his trap door. He must be provided with a small flash-lamp to illuminate the oncoming molecules, and this flash-lamp, since it must give out radiation different in character from that in the vessel, necessarily operates irreversibly. Brillouin shows that however well designed the flash-lamp may be, the entropy it generates always exceeds the decrease due to any segregation of molecules achieved with its aid. Thus there is no net decrease of entropy. Although very few

† L. Brillouin, *J. Appl. Phys.* **22**, 334 (1951).

hypothetical experiments employing fluctuations have been analysed in such detail, it appears most probable that they all fail to violate the second law on account of the necessary entropy generation by the observer who controls the process. There is thus no justification for the view, often glibly repeated, that the second law of thermodynamics is only statistically true, in the sense that microscopic violations repeatedly occur but never violations of any serious magnitude. On the contrary, no evidence has ever been presented that the second law breaks down under any circumstances, and even the entropy law appears to have an almost universal validity, except in such futile experiments as we have discussed above, the removal and reapplication of constraints.

The decrease of availability

So far we have considered the entropy law only in relation to an isolated system. Let us now extend the argument to include systems which are immersed in a bath at constant temperature T_0 and subjected to a constant external pressure P_0. We shall not assume that there is necessarily thermal or mechanical contact between the system and its surroundings, and in the event that the system is enclosed within rigid adiabatic walls the results we shall derive will be identical with those derived for an isolated system. But they will now include as other special cases of importance systems which are in either thermal or mechanical contact, or both, with their surroundings.

In any change of the system from one equilibrium state to another, during which the bath is the sole source of heat supplied to the system, the heat entering from the bath must satisfy (7·2), or, since T_0 is constant,

$$Q \leqslant T_0 \Delta S.$$

We may therefore write for the change in internal energy of the system

$$\Delta U = Q + W \leqslant T_0 \Delta S + W.$$

Also if the only source of work is the external pressure,

$$W = -P_0 \Delta V$$

and

$$\Delta U \leqslant T_0 \Delta S - P_0 \Delta V,$$

or

$$\Delta A \leqslant 0, \quad \text{where} \quad A \equiv U - T_0 S + P_0 V. \tag{7·5}$$

The new quantity, A, is called the *availability* of the system. It will be observed that since T_0 and P_0, rather than T and P, enter in its definition, A is not a property of the system alone, but of the system in given surroundings. The availability is a function which has been more commonly employed by engineers than by physicists, and the name expresses its technically important property of measuring the maximum amount of useful work which can be extracted from a system (e.g. a boiler) during a given change in given surroundings. To prove this we need consider only reversible changes, since it is readily shown from the second law that these enable the greatest amount of work to be performed in a given change. For any infinitesimal change

$$\begin{aligned} dA &= dU - T_0 dS + P_0 dV \\ &= (T - T_0)\, dS - (P - P_0)\, dV, \end{aligned} \tag{7·6}$$

in which T and P are the temperature and pressure within the system.† If the change is reversible, $-T\,dS$ is the heat extracted from the system, which may be most usefully employed by transferring it to an ideal Carnot engine working between the temperatures T and T_0. The work done by the engine will then be $-(T-T_0)\,dS$. Also if the system be allowed to expand reversibly by an amount dV, an additional pressure $(P-P_0)$ must be applied externally, and the useful work of expansion is that performed against the additional pressure; the remaining work done by the system is not useful, being expended in pushing back the atmosphere (or any other passive source of the external pressure P_0). Hence $(P-P_0)\,dV$ is the useful work, and the two work terms add to give $-dA$. Thus the decrease in A is a measure of the maximum useful work available.

This interpretation of A leads us to expect that A takes a minimum value when $T=T_0$ and $P=P_0$, for then the system is in equilibrium with its surroundings and cannot act as a source of useful work. In order to examine this idea we may expand A as a Taylor series in S and V about the point where $T=T_0$ and $P=P_0$, on the assumption that such an expansion is valid:

$$A = A_0 + \frac{1}{2}\left(\frac{\partial T}{\partial S}\right)_V (\Delta S)^2 + \left(\frac{\partial T}{\partial V}\right)_S \Delta S\,\Delta V - \frac{1}{2}\left(\frac{\partial P}{\partial V}\right)_S (\Delta V)^2 + \ldots,$$

the first-order terms vanishing (from (7·6)) at the origin of the expansion. Thus A takes a minimum value if the second-order terms are essentially positive for all ΔS, ΔV, i.e. if

$$(\partial T/\partial S)_V > 0, \tag{7·7}$$

$$(\partial P/\partial V)_S < 0 \tag{7·8}$$

and

$$(\partial T/\partial V)_S^2 < -(\partial T/\partial S)_V (\partial P/\partial V)_S. \tag{7·9}$$

† By pressure within the system we mean that pressure which would have to be applied externally to maintain the system in the same state if it were enclosed in completely flexible and extensible walls. We assume in what follows that P is the same for all parts of the system, but the argument is readily generalized to apply to more complex systems.

The first inequality (7·7) is satisfied if C_V is positive but not infinite. That C_V is positive we saw to be a consequence of the zeroth law; we shall return in a moment to the situation which might arise if C_V were infinite. The second inequality (7·8) is satisfied if the adiabatic compressibility k_S $\left(\equiv -\frac{1}{V}\left(\frac{\partial V}{\partial P}\right)_S\right)$ is positive but not infinite. That k_S should be positive is a mechanical requirement for stability; if it were negative the system would spontaneously collapse, since by doing so it would lessen the pressure needed to keep it in equilibrium. The third inequality (7·9) may be transformed, by means of (6·17) and the fact that $(\partial T/\partial S)_V$ is positive, into the form

$$(\partial P/\partial T)_V (\partial T/\partial V)_S > (\partial P/\partial V)_S.$$

Now $$(\partial P/\partial V)_S = (\partial P/\partial V)_T + (\partial P/\partial T)_V (\partial T/\partial V)_S,$$

and therefore (7·9) implies that

$$(\partial P/\partial V)_T < 0.$$

Thus (7·9) is satisfied if the isothermal compressibility k_T is positive, and this is also a mechanical requirement for stability. Under normal circumstances then, when C_V, k_S and k_T are finite, A takes a minimum value when $T = T_0$ and $P = P_0$, as expected. We may, however, conceive of less usual situations in which some or all of the inequalities become equalities. Two cases may be considered, the first being that in which (7·9) is an equality (k_T infinite) while C_V and k_S are both finite. In this case the second-order terms may be either positive or zero, and the higher order terms in the expansion must take such values as to give A an absolute minimum. This is the situation which arises at the critical point of a liquid-vapour system (see Chapter 8), and it may be shown by a more detailed analysis that $(\partial^2 P/\partial V^2)_T$ must vanish and $(\partial^3 P/\partial V^3)_T$ must be negative. The second case to be considered is that all the inequalities become equalities. Now the third-order terms in the expansion must all vanish and the fourth-order terms must be essentially positive. It may be shown that this imposes more conditions on the form of A than can be simultaneously satisfied, and that therefore C_V and k_S can never become infinite.†

This analysis is valid, however, only on the assumption that a series expansion of A is possible, and we must inquire into the justification for this assumption. Undoubtedly the conclusions reached are correct in the majority of cases, for it is only under exceptional circumstances that C_P and k_T become infinite, and it is still more exceptional to find C_V or k_S infinite. Nevertheless, quite simple examples may be adduced to show the danger of taking for granted the validity of the expansion,

† L. Landau and E. Lifshitz, *Statistical Physics* (Oxford, 1938), p. 102.

and the most elementary of these is a system consisting of two phases in equilibrium, a liquid and solid for instance, and open to the surroundings so that $P = P_0$, $T = T_0$. If P_0 and T_0 are such that both phases may coexist with arbitrary proportions of the material in each phase, it is clear that k_T is infinite, since isothermal reduction of volume merely alters the proportions (e.g. causes more solid to be formed if the solid is

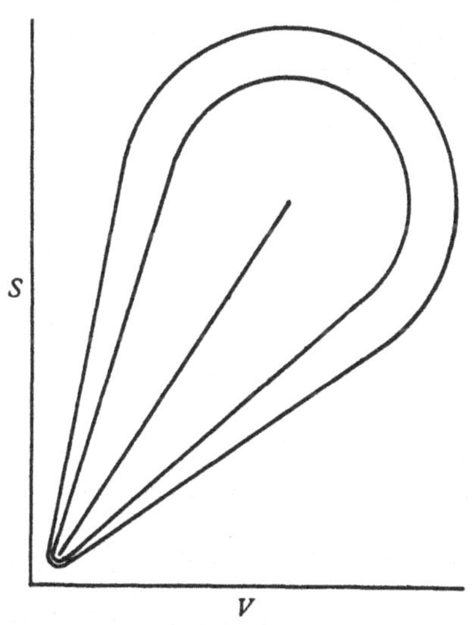

Fig. 23. Contours of availability on the S-V diagram of a two-phase system, when (P_0, T_0) is a point on the phase-equilibrium line. The system is in equilibrium, with varying proportions of the two phases, along the central line.

denser than the liquid) without involving any change of pressure. Such a compression changes S and V without changing the value of A; thus the surface representing A as a function of S and V takes the form shown diagrammatically in fig. 23. So long as both phases are present the surface has a perfectly flat valley, and it is only when the extent of the compression or expansion is such as to eliminate one or other of the phases that the bottom of the valley begins to rise and to reveal that in equilibrium A does indeed take a minimum value. In fact, as remarked in Chapter 2, this is a situation of neutral equilibrium, and no series expansion of A can demonstrate that at the ends of the range of neutrality the surface rises rather than falls. It will be observed that the direction of the bottom of the valley is such that

$$\mathrm{d}S/\mathrm{d}V = (s_l - s_s)/(v_l - v_s),$$

which by Clapeyron's equation (5·12) is just the rate of variation of the melting pressure with temperature, $(\mathrm{d}P/\mathrm{d}T)_{\text{melting}}$. For most substances $(\mathrm{d}P/\mathrm{d}T)_{\text{melting}}$ is neither zero nor infinite, since usually $(s_l - s_s)$ and $(v_l - v_s)$ do not vanish. Under these circumstances both $(\partial^2 A/\partial S^2)_V$ and $(\partial^2 A/\partial V^2)_S$ are non-vanishing, as may be seen from fig. 23, and C_V and k_S remain finite even though C_P and k_T are infinite. There is no reason, however, why under very special conditions $(\mathrm{d}P/\mathrm{d}T)_{\text{melting}}$ should not become either zero or infinite, as illustrated by the points X and Y in fig. 24. It is believed that a point such as X exists on the melting curve of the light isotope of helium, $_2He^3$ (see p. 124), and undoubtedly a point such as Y would exist on the melting

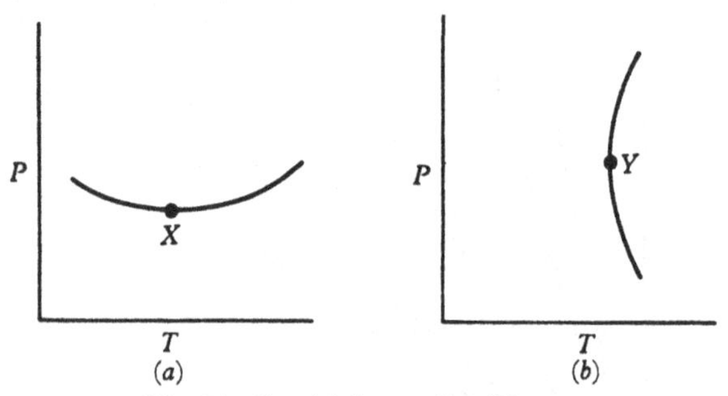

Fig. 24. Special forms of melting curve.

curve of ice were it not for the transformation of ice into another solid modification before this point is reached. At X the valley in fig. 23 has swung round parallel to the V-axis, so that $(\partial^2 A/\partial V^2)_S$ is zero, or k_S infinite, while at Y the valley lies parallel to the S-axis, and C_V is infinite. These examples should make it clear that the finiteness of C_V and k_S is not an absolute thermodynamic requirement, but at the same time exceptions are very rare indeed.†

The conditions for equilibrium

We are now in a position to apply the result expressed by (7·5) to see what it is that characterizes the equilibrium state of a system under given constraints. What we wish to do is to compare various states of the system, of which only one is truly the equilibrium state, and to see how this is distinguished from the rest. As a result of this investigation we may expect to derive a rule for calculating in any given situation the configuration of the equilibrium state. At first sight it might appear a hopeless task, since only in the equilibrium state are we justified in ascribing values to the entropy of the system and to the

† More subtle cases where series expansion of thermodynamic functions must be treated with great caution are examined by L. Tisza, *Phase Transformations in Solids*, ed. Smoluchowski, Mayer and Weyl (Wiley, 1951), ch. 1.

other thermodynamic functions which depend on the entropy. However, by the judicious use of additional constraints we may in imagination convert a variety of non-equilibrium states into equilibrium states, and then inquire into the consequences of removing these constraints. If we are to employ the result expressed by (7·5), the additional constraints which we use must be of such a nature that no sources of work or heat, apart from the surroundings, are involved in their removal. We may, for example, interpose an adiabatic wall between two bodies, or separate two masses of gas by means of a partition; but we must beware of dividing a body into such small pieces that surface effects are introduced or modified. It is difficult to formulate any firm rules for what additional constraints are permissible, but it should become clear in what follows that most examples of physical interest are so simple as to raise few doubts as to their validity.

Let us express the result of (7·5) in language similar to the second formulation of the entropy law:

If the constraints of a system, whose surroundings are maintained at constant temperature and pressure, are altered in any manner which involves no work or heat other than that provided or absorbed by the surroundings, the availability of the system is not thereby increased.

This tells us that when we remove any additional constraints, and allow the system if it wishes to take up a new configuration, the availability of the new configuration cannot be greater than that of the old. Therefore, if we calculate the availability of all conceivable constrained states of a given system, that which has the smallest availability is the equilibrium state of the unconstrained system. For if we remove the additional constraints from this state there is no other configuration to which it can change without making possible, by reapplication of the constraints, an increase of the availability.

A few simple illustrations should make the argument clear. First, two bodies at different temperatures in an enclosure formed of adiabatic walls. If we interpose an adiabatic wall between them so that each is in equilibrium we may calculate the total availability as the sum of the two separate availabilities, whatever the temperatures of the bodies may be. The argument given above now implies that for a given value of the total internal energy the equilibrium state for the bodies in thermal contact is that for which the combined availability of the separated bodies is smallest, which in this case is the same as when the combined entropy is greatest, and this of course is when the temperatures are equal. For if we have the bodies separated by an adiabatic wall when they are in this state, and then remove the wall and replace it again after an interval, however long, the final state of the separated bodies cannot have changed from the original since there is no state of lower availability to which transition is possible. We have thus discovered the equilibrium state of the bodies in contact. In a similar way we may show that the density of a gas in equilibrium is uniform, by separating the gas into regions of varying density by means of removable partitions, and calculating the availability of the non-uniform gas thus artificially produced.

These two examples are trivial, since the results derived either follow from the zeroth law or are implicitly assumed in the development of the subject. But there are many applications of the same method of reasoning where the answer obtained is not trivial but of considerable value. Consider, for instance, a closed vessel containing only a liquid and its vapour, maintained at a constant temperature T_0. What proportion of the material will be in the liquid phase? To solve this problem we clearly wish to apply an additional constraint so that the proportion of liquid, α, may be varied at will without departure from a state of equilibrium, and obviously the way to do this is to interpose a barrier at the surface of the liquid, separating the two phases. Now we can write the combined availability as a function of α, and find for what value of α it takes a minimum value; this is the required equilibrium condition.

In this example we have considered the constraint to take the form of a mechanical barrier, but we might just as well have imagined the evaporation rate of the liquid and condensation rate of the vapour to be reduced to zero, to enable the system to be in metastable equilibrium for any value of α. The real justification for this procedure is perhaps only to be found in a microscopic picture of the processes occurring at the interface between liquid and vapour, from which it is clear that the equilibrium state is only affected to a quite negligible degree by the rate at which it is established. But there is experimental evidence on this point as well; for instance, the rate of evaporation of liquid mercury is highly dependent on the cleanness of the surface, while the equilibrium vapour pressure is not. This device of imagining the rate of progression of a change of state to be reduced to zero is a very versatile way of applying additional constraints, and finds its most important application in chemical thermodynamics. If we wish to find the equilibrium concentrations of the substances taking part in a given chemical reaction, it is convenient to suppose that the system can be 'frozen' into an unreactive state with any values of the concentrations. It may then be treated as an equilibrium state for the purpose of calculating the availability.† Again the fundamental justification for this procedure is to be found in the microscopic viewpoint; in a mixture of reacting gases, for example, at any instant, the number

† The detailed procedure for carrying out such calculations may be found in any text-book of chemical thermodynamics.

of molecules which are actively engaged in interacting with others is so small, even in a fast reaction, that they do not appreciably affect the thermodynamic parameters of the mixture. But there is also powerful *a posteriori* justification in the success which attends the application of thermodynamical reasoning to problems of chemical equilibrium. It will be seen that this argument is only an extension, into the realm of measurably fast chemical reactions, of our early assumption (Chapter 2) that 'metastable' equilibrium may be treated exactly as stable equilibrium, i.e. immeasurably slow reactions may be ignored.

It is now possible to give in one general statement (which must be interpreted in the light of the foregoing discussion) the rule which determines the equilibrium configuration of a system:

For a system, whose sole external sources of heat and work are its surroundings maintained at constant temperature and pressure, of all conceivable configurations that one is stable for which the availability takes a minimum value.

This statement is not the most usual statement of the criteria for equilibrium. We may derive the conventional forms by considering three special types of system:

(1) *An isolated system.* Whether its volume can alter or not, $U+P_0V$ stays constant, and the condition that A shall take a minimum value is the same as the condition that the entropy shall take a maximum value.

(2) *A system of constant volume in thermal contact with a constant-temperature bath.* Here P is not necessarily equal to P_0 when A is minimized (in (7·6) $\mathrm{d}V=0$ so that $P-P_0$ is indeterminate). But P_0V is constant, and when A is minimized $T=T_0$. Thus the condition for equilibrium is that the free energy F ($\equiv U-TS$) shall be minimized, subject to V being constant and T being equal to the external temperature.

(3) *A system acted upon directly by a constant external pressure, in thermal contact with a constant-temperature bath.* If all parts of the system have the same pressure, then, as we have already shown, A takes a minimum value when $T=T$ and $P=P_0$. Thus the condition for equilibrium is that the Gibbs function G ($\equiv U-TS+PV$) shall be minimized, subject to P and T being equal to the external pressure and temperature respectively.

It will be observed that in each of these cases the particular function which takes an external value may be calculated for the complete system by summation over all constituent parts, as pointed out on p. 44. In particular, in case (3), the Gibbs function is defined for the system as a whole because P and T are constant over the system. Now although the temperature takes the same value at all points of a system in equilibrium, however complicated the system may be, it is not necessary that the pressure shall be uniform. For example, the pressure within a liquid drop is greater than that without, on account of surface tension, but a system consisting of the drop and its vapour may still be in equilibrium. In a case like this the Gibbs function of the whole system may be defined, if it is so wished, as the sum of contributions from the different parts; but there is little point in making the effort to define G under these conditions, since it has now no relevance to the criteria for equilibrium. If the system is open to the surroundings, and P takes different values in different parts, the conditions are not those of case (3) above, and we must revert to our original statement of the criterion for equilibrium, that the availability A takes a minimum value. Since the pressure and temperature, P_0 and T_0, which enter into A are by definition constant, A is always an additive quantity for a system in equilibrium. We shall return at the end of this chapter to a more detailed analysis of the equilibrium of a liquid drop with its vapour.

From this discussion it will be clear that the choice of criterion for distinguishing the equilibrium state of a system depends on the nature of the constraints to which the system is subjected, and it is of interest to demonstrate the relation between the different criteria by a specific calculation. We shall consider the equilibrium between a liquid and its vapour, under two different constraints: first, with the vessel immersed in a constant-temperature bath, and open to a constant external pressure; and secondly, with the vessel closed and thermally isolated. In the first case, let us suppose that in equilibrium there are masses α of liquid and $(1-\alpha)$ of vapour. The Gibbs function of the whole system may be written in the form

$$G = \alpha g_l + (1-\alpha) g_v, \tag{7·10}$$

in which g_l and g_v are the Gibbs functions per unit mass of liquid and vapour. In equilibrium,

$$\left(\frac{\partial G}{\partial \alpha}\right)_{P,T} = g_l - g_v = 0. \tag{7·11}$$

Since the pressure and temperature remain constant, g_l and g_v are independent of α. We see therefore that the condition for equilibrium is that the Gibbs functions per unit mass shall be the same in both phases.

Turning now to the isolated vessel, we may no longer assume g_l and g_v to be independent of α, since varying α changes the temperature and pressure in the vessel. There are, however, certain invariants of the system, its total volume V, and its internal energy U; moreover, in equilibrium the entropy is maximized and therefore stationary with respect to small variations of α. We may thus write

$$V = \alpha v_l + (1-\alpha)\, v_v,$$

or
$$\mathrm{d}V = \alpha\,\mathrm{d}v_l + (1-\alpha)\,\mathrm{d}v_v + (v_l - v_v)\,\mathrm{d}\alpha = 0; \tag{7·12}$$

$$U = \alpha u_l + (1-\alpha)\, u_v,$$

or
$$\mathrm{d}U = \alpha\,\mathrm{d}u_l + (1-\alpha)\,\mathrm{d}u_v + (u_l - u_v)\,\mathrm{d}\alpha = 0; \tag{7·13}$$

$$S = \alpha s_l + (1-\alpha)\, s_v,$$

or
$$\mathrm{d}S = \alpha\,\mathrm{d}s_l + (1-\alpha)\,\mathrm{d}s_v + (s_l - s_v)\,\mathrm{d}\alpha = 0. \tag{7·14}$$

Now multiply (7·12) by P, the internal pressure, and (7·14) by T, the internal temperature, both of which will have the same value in both phases. Then

$$\begin{aligned}\mathrm{d}U + P\,\mathrm{d}V - T\,\mathrm{d}S \\ &= \alpha(\mathrm{d}u_l + P\,\mathrm{d}v_l - T\,\mathrm{d}s_l) + (1-\alpha)\,(\mathrm{d}u_v + P\,\mathrm{d}v_v - T\,\mathrm{d}s_v) \\ &\quad + (u_l + Pv_l - Ts_l)\,\mathrm{d}\alpha - (u_v + Pv_v - Ts_v)\,\mathrm{d}\alpha = 0.\end{aligned}$$

But from the fundamental equation of the second law the coefficients of α and $(1-\alpha)$ vanish, while the remaining terms are $g_l\,\mathrm{d}\alpha$ and $g_v\,\mathrm{d}\alpha$. We arrive therefore at the same equilibrium condition as before, that $g_l = g_v$. The reader may verify by analogous arguments that this result is also obtained with a thermally isolated vessel open to the external pressure (S maximized with P and H constant) and with a closed vessel in thermal contact with a constant-temperature bath (F minimized with V and T constant). The liquid and vapour cannot be in equilibrium unless the Gibbs function per unit mass takes the same value in both phases. This result forms the basis of the theory of phase equilibrium, to which the following chapters are devoted.

We now return to the problem raised on p. 108, and discuss in more detail the equilibrium between a small liquid drop and its vapour, when the latter is maintained at constant temperature T_0 and pressure P_0; on account of surface tension the pressure within the drop is $P_0 + 2\sigma/r$, where σ is the surface tension and r the radius of the drop. The equilibrium state is characterized as that for which the availability of the system is stationary with respect to small variations of r. It is convenient to express the availability of the whole system as a sum of three terms, contributed by the vapour, the bulk of the liquid, and the surface of the drop. So far as the vapour is concerned its availability A_1 may be equated to its Gibbs function

$$A_1(T_0, P_0) = m_v(u_v - T_0 s_v + P_0 v_v) = m_v g_v(T_0, P_0),$$

since T_0 and P_0 are the temperature and pressure, both external to the system and in the vapour itself. On the other hand, the equating of availability and Gibbs function is not strictly possible for the liquid within the drop. For its availability per unit mass is $(u_l - T_0 s_l + P_0 v_l)$, in which u_l, s_l and v_l are to be calculated not at a pressure P_0 but at a pressure $P_0 + 2\sigma/r$. If, however, we make the reasonable assumption that the liquid is in effect incompressible, it follows that $u_l - T_0 s_l$ and v_l are independent of pressure, and we may then write for the liquid

$$A_2(T_0, P_0) = m_l g_l(T_0, P_0).\dagger$$

Finally, we have for the surface term

$$A_3(T_0, P_0) = 4\pi r^2(\lambda - T_0\eta) = 4\pi r^2\sigma(T_0) \quad \text{from (6·57)}$$

Thus for the whole system of mass m

$$\begin{aligned} A(T_0, P_0) &= A_1 + A_2 + A_3 \\ &= (m - \tfrac{4}{3}\pi r^3\rho_l)\, g_v(T_0, P_0) + \tfrac{4}{3}\pi r^3\rho_l\, g_l(T_0, P_0) + 4\pi r^2\sigma(T_0), \end{aligned}$$

where ρ_l is the density of the liquid. Then the equilibrium condition $(\partial A/\partial r)_{T_0, P_0} = 0$ leads to the equation

$$g_v(T_0, P_0) - g_l(T_0, P_0) = \frac{2\sigma(T_0)}{r\rho_l}. \tag{7·15}$$

From this equation it may be seen that if the drop is large, so that the right-hand side is small, P_0 and T_0 must be adjusted in equilibrium so that $g_v = g_l$, and this is the result we obtained before. If r is small the equilibrium relation between P_0 and T_0 is altered, but the new vapour pressure at temperature T_0 is readily expressed in terms of the vapour pressure P_b of the bulk liquid (i.e. for the case $r = \infty$). For we have that

$$g_v(T_0, P_b) = g_l(T_0, P_b) = g_0, \quad \text{say;}$$

also
$$(\partial g_v/\partial P)_{T_0} = v_v = RT_0/(MP),$$

where M is the molecular weight of the vapour, and R is the gas constant per mole; and

$$(\partial g_l/\partial P)_{T_0} = v_l = 1/\rho_l.$$

Hence
$$g_v(T_0, P_0) = g_0 + \frac{RT_0}{M}\log\left(\frac{P_0}{P_b}\right),$$

and
$$g_l(T_0, P_0) = g_0 + \frac{P_0 - P_b}{\rho_l},$$

† It will be instructive for the reader to work out how this expression and the final result are to be modified if the compressibility of the liquid is not negligible.

so that (7·15) may be written

$$\frac{\rho_l RT_0}{M}\log\left(\frac{P_0}{P_b}\right)-(P_0-P_b)=\frac{2\sigma}{r}. \qquad (7\cdot16)$$

If the second term on the left-hand side be neglected, we arrive at Kelvin's formula for the vapour pressure of a drop,

$$P_0=P_b\exp\left\{\frac{2\sigma M}{\rho_l rRT_0}\right\}. \qquad (7\cdot17)$$

The neglect of (P_0-P_b) in (7·16) may be justified by expanding the exponential in (7·17) as far as its second term, whence it may be deduced that (P_0-P_b) is smaller than $2\sigma/r$ by a factor ρ_v/ρ_l, ρ_v being the density of the vapour. Thus (7·17) is a good approximation under most circumstances.

It may be noted that the equilibrium of a drop with its vapour is unstable, for if the drop grows a little its equilibrium vapour pressure decreases, and if it is kept in contact with vapour at constant pressure it will continue to grow. Similarly, if it starts to evaporate it will continue evaporating until it disappears. This means, as may be verified by direct calculation, that under isothermal conditions the availability of the system (drop + vapour) has a *maximum*, not a minimum, when the pressure is that given by (7·17). In practice, therefore, the continued coexistence of droplets and vapour will not be observed if the vapour pressure is maintained constant. On the other hand, if the system is enclosed so that the total volume remains constant, it is found, as the reader may verify, that (7·17) gives the condition for F to take an extremal value, and if the drop is not too small the extremum may be a minimum. It is possible in an enclosed system for droplets to exist in stable equilibrium, but only if they are larger than a certain critical size. We shall make use of this result in Chapter 9.

CHAPTER 8

PHASE EQUILIBRIUM

In this chapter we shall consider applications of the result derived in the last chapter that (provided surface effects are negligible) two phases of the same substance can only coexist in equilibrium if they possess the same Gibbs function per unit mass. The examples we shall treat are the equilibrium of the solid, liquid and vapour phases of a substance at various pressures and temperatures, the phase diagrams of the two isotopes of helium, and the equilibrium between the normal and superconducting phases of a metal, for which we must introduce the magnetic field as an additional parameter. These examples should be sufficient to illustrate the methods involved in the treatment of phase equilibrium. No essentially new ideas are involved in the extension of these methods to more complex systems, such as solids which may exhibit many allotropic forms (ice, sulphur). Further, a full understanding of these examples should enable the reader, if he wishes, to grasp the thermodynamic theory of chemical reactions and of the phase diagrams of alloys with little difficulty, as the methods there employed are fundamentally of the same nature.

The phase diagram of a simple substance

We consider a simple substance which under suitable conditions exhibits three modifications, solid, liquid and vapour; by 'simple' we imply no more than that only these three modifications exist, at any rate under the conditions of temperature and pressure with which we are concerned. In order to determine the circumstances under which any of these phases can coexist it is convenient to study the variation with temperature and pressure of the Gibbs function g (in what follows it will be taken for granted that we are considering unit mass of each phase). Since for any change

$$dg = -s\,dT + v\,dP,$$

therefore
$$\left(\frac{\partial g}{\partial T}\right)_P = -s, \quad \left(\frac{\partial g}{\partial P}\right)_T = v. \tag{8·1}$$

If then we exhibit g as a function of P and T, the surface representing g (*g-surface*) will always slope upwards in the direction of increasing pressure, the gradient being small and nearly constant for a liquid or solid, and steep for a gas, decreasing as P increases. The slope in the direction of increasing temperature will depend on the choice of the arbitrary constant in s, but if, as is usual, this is chosen so that s is positive at all temperatures the slope of the surface will always be negative, increasingly so as T increases since c_p and therefore $(\partial s/\partial T)_P$ are always positive. Thus the surface has everywhere a negative curvature,† but the detailed shape may be expected to vary considerably for different phases of the same substance.

Let us suppose that we can construct one of these g-surfaces for each phase, irrespective of the question of whether the phase is observable at all pressures and temperatures. We then have three surfaces of which any two in general intersect along a line, the three lines so constructed meeting at the point at which all three surfaces intersect

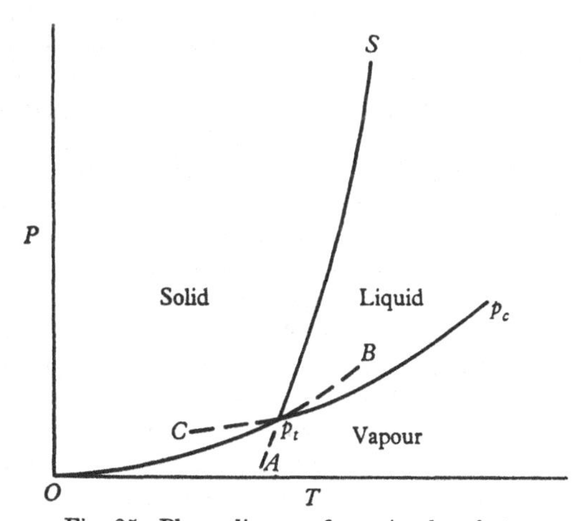

Fig. 25. Phase diagram for a simple substance.

one another. The projection of these three lines onto the P-T plane will give a diagram of the type shown in fig. 25. Along these lines the two phases corresponding to the intersecting surfaces may coexist in equilibrium, since they have the same value of g, and at p_t all three phases may coexist. The point p_t is thus the triple point of the substance, and our experimental knowledge of the behaviour of normal substances enables us to interpret Op_t as the sublimation curve, along which solid and vapour are in equilibrium, p_tp_c as the vapour pressure curve, along which liquid and vapour are in equilibrium, and p_tS as the melting curve, along which liquid and solid are in equilibrium. At points removed from these lines one of the g-surfaces lies lower than

† Strictly this statement is true only if the set of quadratic terms in the expansion of g as a power series in P and T is essentially negative. It may be left as an exercise for the reader to demonstrate that this is so if $c_V > 0$.

the others and this determines which phase is stable, since the equilibrium state of the system at given P and T is that for which the Gibbs function is minimized. It is clear that in the region below Op_tp_c the vapour phase is stable, for it possesses the largest volume and therefore the Gibbs function falls most steeply as the pressure is reduced. It is not so obvious that the region to the left of Op_tS corresponds to the stable existence of the solid and the remaining region to the stable existence of the liquid, and here we must rely on experimental observation to ascribe the regions correctly. As the curves are drawn in the figure it will be seen that in crossing the line p_tS vertically upwards we move from a region of liquid stability to one of solid stability; this implies that $(\partial g_l/\partial P)_T > (\partial g_s/\partial P)_T$, i.e. that $v_l > v_s$, or that the substance expands on melting. For substances such as water or bismuth which contract on melting, p_tS will have the opposite slope. If we move across an equilibrium line horizontally (P constant, T increasing) it is obvious that the phase which is stable on the high-temperature side must have the greater negative slope $-(\partial g/\partial T)_P$, that is, by (8·1), the higher entropy. Therefore in a reversible transition from the low-temperature to the high-temperature phase heat must always be absorbed, and the latent heat is in consequence always positive.

It might be argued that since only one phase is stable except along the equilibrium lines it is not justifiable to imagine the g-surfaces to extend over the whole plane for each phase, and indeed it is quite unnecessary to make this supposition. We might equally well have started constructing the surface from three points, one in the middle of each region, making use of experimental information to decide which phase to consider in each region; the intersections could then be found without supposing the surfaces to be capable of crossing one another. In fact, for many substances, particularly if they are highly purified, it is possible to observe the metastable existence of a phase in a region where it should have made a transition to another phase. The most striking examples of this behaviour are the supercooling of vapours, which may stay uncondensed at pressures four times or more higher than the equilibrium vapour pressure (the Wilson cloud chamber owes its operation to this fact), and the supercooling of liquids without solidification. For instance, small drops of water in a cloud may be cooled to $-40°$ C. before they freeze. The converse effects may also be demonstrated, as, for example, the superheating of liquids, which is often manifested as 'bumping', and which is the basis of the operation of the bubble chamber used to detect high energy particles. There is therefore no real obstacle to imagining each surface to continue beyond its intersection with another, and the foregoing argument could be dismissed as trivial except for the fact that much confusion has been generated in discussions of higher-order transitions (see Chapter 9) by a too-facile assumption that g-surfaces are always continuable beyond an equilibrium line. Before leaving the question of the metastable persistence of phases, it may be pointed out that the equilibrium lines may be imagined prolonged through the triple point, as p_tA, p_tB and p_tC in fig. 24. Along p_tA, for example, $g_l = g_s$, but both are greater than g_v. There is no reason in principle why a superheated liquid should not be made to solidify by crossing the curve p_tA, though the author is not aware of any observation of this phenomenon. On the other hand, the coexistence of a supercooled liquid with its vapour along the line p_tC is a commonplace of meteorology; the fact that the vapour pressure of the supercooled liquid is higher than that of the solid at the same temperature (as shown in fig. 25) means that any ice crystal in a cloud containing supercooled water drops tends to grow rapidly, since the latter maintain a high vapour pressure.

It is a result of general validity that the vapour pressure of a supercooled liquid is higher than that of the solid at the same temperature. For if this were not so the line p_tC would lie in the vapour region below Op_tp_c; but in this region g_v is lower than either g_l or g_s, while at the point C, $g_v = g_l$. It follows then that around a triple point there is only one metastable line in each region, and that therefore the angles between neighbouring equilibrium lines are never greater than 180°.

Clapeyron's equation

Along any equilibrium line separating two phases in the P-T diagram the Gibbs functions are equal, $g_1 = g_2$. The suffix 1 denotes that phase which is stable on the low-temperature side of the equilibrium line. For small variations of pressure and temperature, δP and δT, which alter the state of the system to a neighbouring state still on the equilibrium line, the variation of g_1 and g_2 must therefore be equal, so that

$$\left(\frac{\partial g_1}{\partial T}\right)_P \delta T + \left(\frac{\partial g_1}{\partial P}\right)_T \delta P = \left(\frac{\partial g_2}{\partial T}\right)_P \delta T + \left(\frac{\partial g_2}{\partial P}\right)_T \delta P,$$

or

$$(s_2 - s_1)\,\delta T = (v_2 - v_1)\,\delta P \quad \text{from (8·1)}.$$

Now the ratio $\delta P/\delta T$ tends, as $\delta T \to 0$, to the slope $\mathrm{d}P/\mathrm{d}T$ of the equilibrium line, so that

$$\frac{\mathrm{d}P}{\mathrm{d}T} = \frac{s_2 - s_1}{v_2 - v_1} = \frac{l}{T(v_2 - v_1)}, \tag{8·2}$$

where l is the latent heat of the transition, $T(s_2 - s_1)$, per unit mass. This is Clapeyron's equation which we derived (5·12) in Chapter 5 by a different argument. From this point of view it is clear that

Clapeyron's equation is simply an expression of the fact that along a transition line the g-surface has a sharp crease. If one describes the slope of the g-surface by the two-dimensional vector grad g, at the transition line there is a discontinuity $\Delta(\text{grad}\, g)$ in the slope, and Clapeyron's equation states that the vector $\Delta(\text{grad}\, g)$ is directed normal to the transition line.

Since l cannot be negative it may be seen from equation (8·2), as we deduced by a geometrical argument in the last section, that a solid which expands on melting ($v_2 > v_1$) has a positive pressure coefficient of the melting temperature, and one which contracts has a negative coefficient. The comparison, by James and William Thomson in 1849–50, of the actual pressure variation of the melting-point of ice with that predicted by Clapeyron's equation is of historical interest as being perhaps the first successful application of thermodynamics to a physical problem, and the success of this simple test undoubtedly contributed largely to the spirit of confidence which underlay and encouraged the rapid development of the subject.

Liquid-vapour equilibrium and the critical point

Along the curve $p_t p_c$ in fig. 25 the g-surfaces for the liquid and vapour intersect and the two phases are in equilibrium. This line does not, however, continue indefinitely, for at the critical point p_c the liquid and vapour become indistinguishable. The critical phenomenon is clearly illustrated by the isotherms of a typical liquid-vapour system shown in fig. 26. If the system is in the vapour phase at a temperature below the critical temperature T_c, and is compressed isothermally, a stage is reached, as at C, when liquid begins to form and the system becomes inhomogeneous. Subsequent compression, at constant pressure, eventually leads to the state C' where the vessel is wholly filled with liquid, which is far less compressible than the gas phase. In the inhomogeneous region CC' the isothermal compressibility, defined as $-\frac{1}{V}\left(\frac{\partial V}{\partial P}\right)_T$, is of course infinite. At the critical temperature the compressibility decreases at first as the volume is reduced, and then rises to infinity at the horizontal point of inflexion p_c, which corresponds to p_c in fig. 25; further compression reduces the compressibility steadily. At no point is there a separation into two phases. At temperatures well above T_c the compressibility decreases monotonically as the volume is reduced, so that although at large volumes the substance may be regarded as vapour-like, and at low volumes as liquid-like, there is no point at which any transition from one phase to another may be said to have occurred. It is therefore possible, by travelling around the point p_c in fig. 25, to make a transition from what is undoubtedly the vapour phase to the equally undoubted liquid phase, without any abrupt change in the properties of the system. This means that the g-surfaces for the liquid and the vapour must both be parts of the same surface, which, however, is not of a simple form, but intersects itself along the line $p_t p_c$, though not beyond p_c. In order to obtain a picture of the form of this surface it is

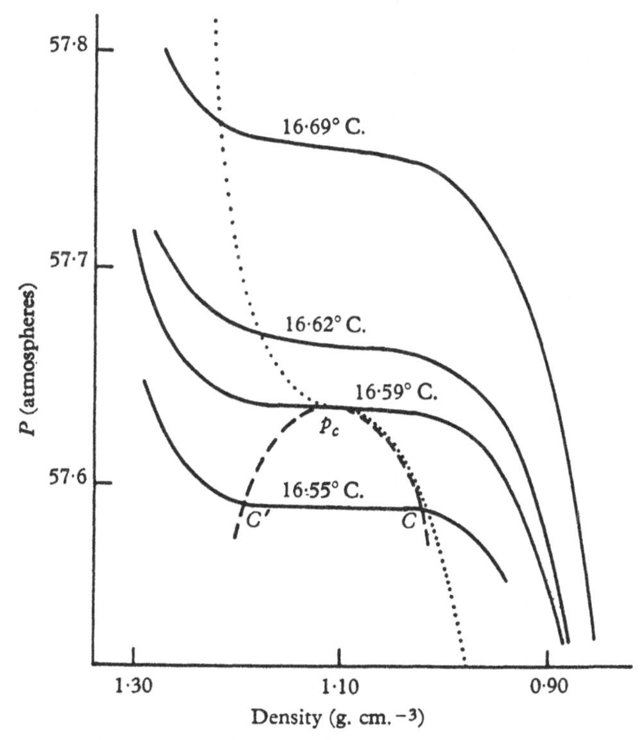

Fig. 26. Isotherms of xenon near the critical point (H. W. Habgood and W. G. Schneider, *Canad. J. Chem.* **32**, 98, 1954). The broken line marks the region of coexistent phases, and the dotted line is the critical isotherm according to the van der Waals equation.

convenient to make use of the equation of state proposed by van der Waals (p. 75 footnote) to express the continuity of the liquid and vapour phases. The critical isotherm of a van der Waals gas and a typical isotherm for a temperature below T_c are shown in fig. 27, together with the conjectured horizontal line CC' which corresponds to the mixture of phases. We shall for the moment ignore this line and suppose that the whole of the curve for a temperature less than T_c is experimentally realizable, in spite of the fact that in the central region the compressibility is negative and the hypothetical homogeneous phase is intrinsically unstable.

We now use this isotherm to calculate the variation of g with pressure at constant temperature, since $(\partial g/\partial P)_T = v$, and therefore

$$g(P,T) = g(P_0,T) + \int_{P_0}^{P} v\,\mathrm{d}P,$$

the integral being taken along the isotherm from P_0 to P. Let us take the point D to correspond to the initial pressure P_0. The first stage of

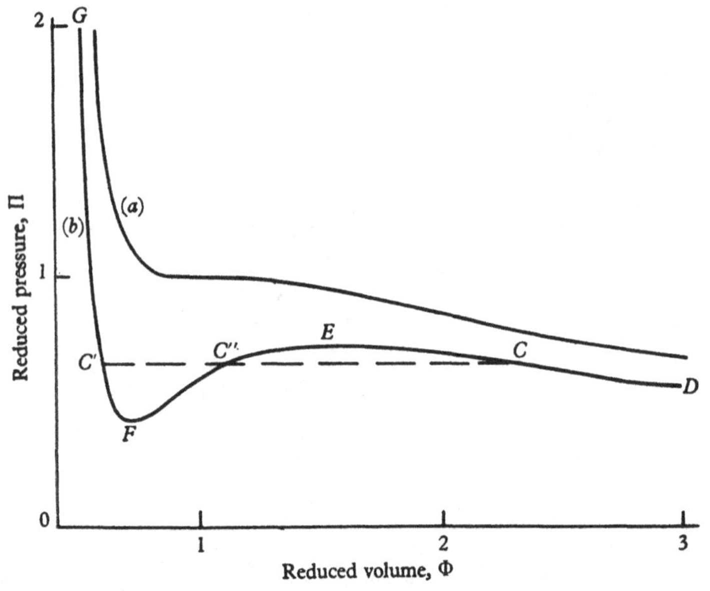

Fig. 27. Van der Waals isotherms: (*a*) $T = T_c$, (*b*) $T = 0{\cdot}9T_c$.

the integration, from D to E, is straightforward, and leads to the curve DE in fig. 28. Between E and F the contribution of the integral is negative, and g falls, but at the same time P also falls, so that $(\partial g/\partial P)_T$ is still positive (as it must be, being equal to v). The curve for g shows a cusp at E, the tangents to the two curves becoming coincident at E. Similarly, there is another cusp at F, and then the curve proceeds steadily upwards towards and beyond G. Clearly the portion DE corresponds to the vapour phase and the portion FG to the liquid phase, and equilibrium of the two phases is possible at the pressure P_e where they intersect. The difference in gradient between

DE and FG at this point represents the volume difference between the two phases. Since g takes the same value at C and C', it is clear that the line CC' must be drawn at such a pressure P_e that $\int_C^{C'} v\,dP = 0$, that is, the areas $C'FC''$ and CEC'' must be equal.† Now this argument, and that given in the footnote, both assume that the whole of the van der Waals isotherm is physically realizable as a succession of equilibrium states of the system. As this is not true it is very doubtful what validity the arguments possess. From a logical point of view they are worthless, but still it is probable that they lead to the correct conclusion. It would require a much more detailed investigation of the model to justify this result completely.

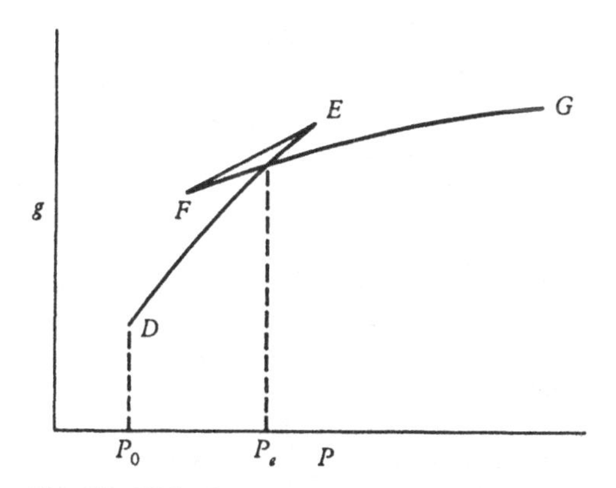

Fig. 28. Gibbs function obtained by integrating the van der Waals equation.

Returning now to fig. 28 we can see that as the temperature is raised and $v_v - v_l$ diminishes, the two branches DE and FG intersect more and more nearly tangentially, and the cusped region becomes steadily smaller, until at the critical temperature the curve degenerates into a single continuous curve. Just at the critical temperature the gradient of the curve, which is equal to v, is everywhere continuous, but the curvature $(\partial v/\partial P)_T$ becomes momentarily infinite at the critical pressure, since at this point the van der Waals gas is infinitely compressible. Above T_c the curves for g are everywhere continuous in all their

† An alternative argument leading to the same conclusion runs as follows: imagine the substance to be taken through the cycle $CEC''FC'C''C$ reversibly. This cycle is completely isothermal, so that its efficiency must be zero, i.e. no work must be done in the cycle. Hence the area of the cycle must vanish, a condition satisfied by making the two areas $C'FC''$ and CEC'' equal.

derivatives. A sketch of the surface for g is shown in fig. 29. The top portion, between the two lines of cusps, is the completely unrealizable region; a part of the regions between the line of intersection and the lines of cusps may be realized by superheating of the liquid or supercooling of the vapour.

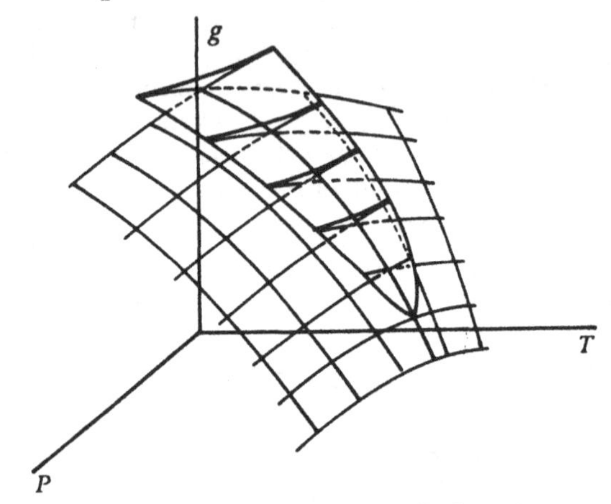

Fig. 29. Gibbs surface for the van der Waals gas in vicinity of the critical point.

Along with anomalous compressibility in the neighbourhood of the critical point the specific heat also shows certain anomalies, which we may exhibit qualitatively with the help of van der Waals's equation. It will be recalled (see p. 60) that according to this equation $(\partial c_V/\partial v)_T = 0$, so that if, as is a good approximation for many gases, $(\partial c_V/\partial T)_V$ is effectively zero, c_V is the same for all states of the gas or liquid. This is certainly not true in practice, and illustrates one of the limitations of van der Waals's equation, but, nevertheless, it serves to show that we need not expect any markedly unusual behaviour in c_V near the critical point. On the other hand, as the isothermal compressibility k_T goes to infinity at the critical point, so does $c_P - c_V$, since, as follows easily from (6·9),

$$c_P - c_V = vTk_T\left(\frac{\partial P}{\partial T}\right)_v^2. \tag{8·3}$$

A quantitative calculation from van der Waals's equation is straightforward. If we take the critical point as our origin of coordinates, and work in reduced coordinates, writing

$$x \equiv \frac{v - v_c}{v_c}, \quad y \equiv \frac{P - P_c}{P_c}, \quad z \equiv \frac{T - T_c}{T_c},$$

the equation takes the form (when all powers of x above x^3 are neglected),

$$3x^3+(2+3x)\,y-8z=0. \tag{8·4}$$

Equation (6·9) may be rewritten in the form

$$c_P-c_V=\frac{P_c v_c T}{T_c^2}\left(\frac{\partial y}{\partial z}\right)_x\left(\frac{\partial x}{\partial z}\right)_y,$$

and near the critical point, where T may be put equal to T_c, and $P_c v_c=\frac{3}{8}rT_c$,

$$c_P-c_V=\frac{8r}{(2+3x)\,(y+3x^2)}. \tag{8·5}$$

From this result it will be seen that if the pressure is maintained at the critical value ($y=0$) and the volume altered by changing the temperature, c_P-c_V, and hence c_P, approach infinity as const./x^2. Since, from (8·4), $z\propto x^3$ when $y=0$, we see that at the critical pressure c_P approaches infinity as const./$(T_c-T)^{\frac{2}{3}}$. If the volume is maintained at v_c ($x=0$), c_P approaches infinity as const./(T_c-T).

We may further use this model to demonstrate that although c_P and k_T become infinite at the critical point, the adiabatic compressibility remains finite as asserted in Chapter 7. For, from (8·5), as the critical point is approached γ ($\equiv c_P/c_V$) tends to infinity,

$$\gamma\to\frac{8r/c_V}{(2+3x)\,(y+3x^2)},$$

while the isothermal compressibility tends to the form $\dfrac{1}{P_c}\dfrac{2+3x}{3(y+3x^2)}$.

Therefore

$$k_S=k_T/\gamma\to\frac{c_V}{24rP_c}(2+3x)^2,$$

which remains finite when $x=0$.

It is important to remember, however, that although the van der Waals model illustrates qualitatively the thermodynamic behaviour near the critical point, it is very far from being quantitatively exact. The isotherms of a real substance in the vicinity of the critical point tend to be much flatter than those predicted by van der Waals's equation, as may be seen in fig. 26. It seems that not only $(\partial^2P/\partial V^2)_T$, but $(\partial^3P/\partial V^3)_T$ and $(\partial^4P/\partial V^4)_T$ also vanish at the critical point, and it is likely that the behaviour cannot be adequately represented, as in van der Waals's and similar equations, by functions which are capable of expansion as a Taylor series involving values of the derivatives at the critical point itself. But this is too complex a problem to be entered into here.

Solid-liquid equilibrium

Let us now examine the question whether the transition line between the solid and liquid phases ($p_t S$ in fig. 25) continues indefinitely, or whether it also terminates abruptly in a critical point. It may be said at the outset that there are weighty theoretical arguments against the possibility of a critical point. For if such a point existed it would be possible, by going around it, to make a continuous transition from the liquid to the solid phase. Now the properties of normal liquids are strictly isotropic; they possess no crystalline structure which singles out any one direction as different from another, while true solids (excluding glasses and similar amorphous phases) possess non-spherical symmetries which are characteristic of the regular arrangement of their molecules in a crystalline lattice. In order to go from the liquid to the crystalline phase, therefore, it is necessary to make a change of the symmetry properties, and this is of necessity a discontinuous process. The symmetry properties of a lattice are describable in terms of certain geometrical operations, such as translation or reflexion, which displace every atom on to another identical atom and so leave the lattice unaltered. A given phase either possesses or does not possess any given symmetry property, and thus no continuous transition is possible and no critical point can exist.

This is the theoretical argument, which has appeared to some to be a little too straightforward to be absolutely convincing. There is little doubt, however, that the experimental evidence all points strongly to the truth of its conclusion. Of this evidence the most complete is that of Simon and his coworkers on solid and liquid helium. The choice of helium as a suitable substance for extensive investigation was governed by the following considerations. According to the *law of corresponding states*, the phase diagrams of most simple substances are very similar in form and scale if they are plotted in reduced coordinates, P/P_c, V/V_c and T/T_c. There are, of course, differences in detail, but on the whole the law is well obeyed. It is therefore desirable to study a substance for which the highest values of P/P_c and T/T_c are attainable, and this is achieved by using helium, whose critical pressure and temperature are 2·26 atmospheres and 5·2° K. respectively. Since the melting curve of helium can be followed up to pressures of several thousand atmospheres, it follows that values of P/P_c of more than one thousand may be attained, far more than is possible with any other substance.

Apart from the lowest temperature region, in which helium behaves unlike any other substance (this will be discussed more fully in the next section), the melting curve takes a form which closely resembles that of many other substances, in that the melting pressure varies with temperature according to the law $P/a=(T/T_0)^c-1$, in which a and T_0 are constants for a given substance and c is an exponent which varies somewhat from one substance to another, but usually lies between 1·5 and 2. The melting curve for helium is shown in fig. 30. There is thus no indication of a critical point for the liquid-solid transition even at temperatures eight times as high as that of the liquid-vapour critical point. It is interesting to note that if the same law for the melting

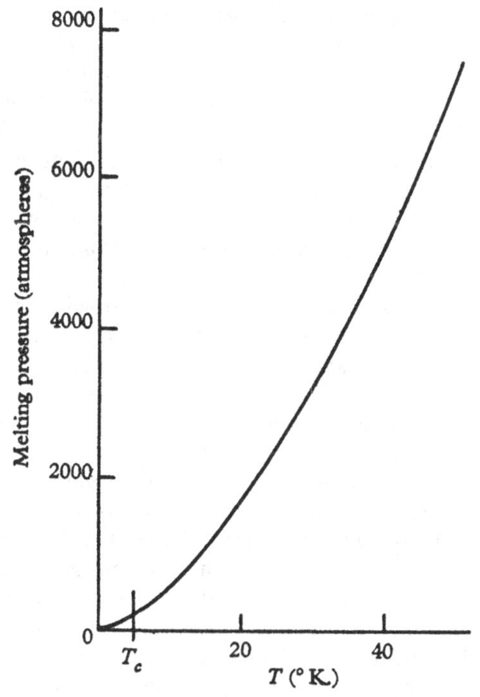

Fig. 30. Melting curve of helium (F. A. Holland, J. A. W. Huggill and G. O. Jones, *Proc. Roy. Soc.* A, **207**, 268, 1951).

pressure holds beyond the range of existing data, helium could be solidified at room temperature by application of a pressure of 110,000 atmospheres, not so much higher than the highest pressures produced by Bridgman. There are, however, technical reasons limiting the pressure which can be applied to helium at a much lower value.

From the point of view of determining whether a liquid-solid critical point exists, an even more instructive set of measurements is that of the entropies of the liquid and solid phases, which have been computed from specific heat measurements at high pressures. It is found that the entropy difference between liquid and solid helium increases slightly as the melting pressure is increased. Since at a critical point the two phases, being identical, must have the same entropy, the experiments make it highly probable that no critical point will be found.†

The phase diagram of helium

Naturally occurring helium consists almost entirely of $_2He^4$, the isotope of mass 4 (two protons and two neutrons comprising the nucleus), with an admixture of one part in 10^5–10^6 of $_2He^3$, the light isotope of mass 3 (two protons and one neutron). Until recently little was known of the properties of the latter, but with the development of methods of separating the two isotopes and, still more, with the production on a comparatively large scale of $_2He^3$ in nuclear reactors,‡ it has become possible to carry out experiments on the thermodynamic and other properties of this very interesting substance, and even to use it as a liquid bath in cryostats. Since it behaves in a rather more simple manner than $_2He^4$ we shall discuss it first.

Both isotopes have the property, unlike any other substance, that they may remain liquid down to the lowest temperatures. The reason for this, which is essentially quantum-mechanical, need not concern us here,§ since we shall limit our discussion to thermodynamical aspects of the phase diagram. The diagram for $_2He^3$ is shown in fig. 31, from which it will be seen that there is no triple point.

The form of the melting curve is unique among liquid-solid transitions. At all temperatures the solid is the denser phase, but whereas above 0·32° K. the entropy of the liquid exceeds that of the solid, as is usual, below 0·32° K. the reverse is true. There is a range of temperatures and pressures within which the liquid is the low temperature phase and the solid the high temperature phase. If a vessel containing $_2He^3$ is maintained at a pressure of 30 atmospheres and warmed up from a very low temperature, it starts as a liquid, freezes at 0·18° K., and melts once more at 0·49° K. It is worth noting in passing that solid iron behaves somewhat analogously, as may be seen in fig. 41; on heating it makes a transition from the α-modification to the γ-modification, and at a higher temperature reverts to the α-modification. The melting curve of $_2He^3$ has been studied down to 0·06° K.‖ and is

† For details of these experiments see J. S. Dugdale and F. E. Simon, *Proc. Roy. Soc.* A, **218**, 291 (1953).

‡ The reaction $_3Li^6 + _0n^1 \rightarrow _2He^4 + _1H^3$, creates tritium $_1H^3$, which decays by β-emission with a half-life of twelve years to $_2He^3$.

§ See W. H. Keesom, *Helium* (Elsevier, 1942), p. 332.

‖ D. O. Edwards, J. L. Baum, D. F. Brewer, J. G. Daunt and A. S. McWilliams, *Proc. 7th Int. Conf. on Low Temperature Physics* (Toronto, 1961), p. 610.

still rising steadily. Ultimately it must become level, since according to the third law $S_s - S_l$ must vanish at the absolute zero, but the temperature at which this occurs may prove to be very low indeed.

The melting curve of $_2He^4$ is more normal, though again there is no triple point, and a pressure of at least 25 atmospheres is needed to produce the solid. Of particular interest in the phase diagram (fig. 32) is the line separating two different forms of the liquid phase.

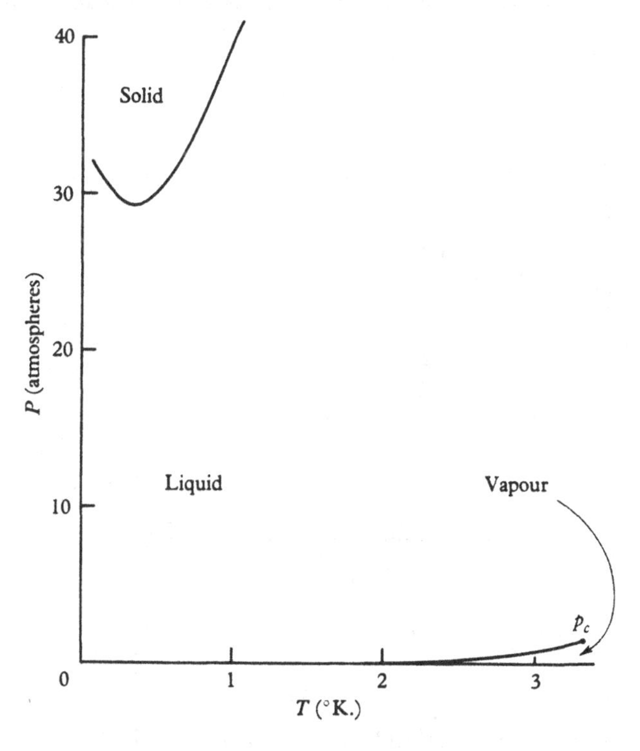

Fig. 31. Phase diagram for $_2He^3$.

The higher-temperature form, He I, shows no markedly unusual properties; but as the temperature is lowered a sharp transition to a new form, He II, occurs, which possesses such remarkable and unique properties of high heat conductivity and low viscosity as have earned it the title 'superfluid'. Undoubtedly, the great fascination of helium is due to its transport properties and hydrodynamic peculiarities, but its thermodynamic behaviour is not devoid of interest, since it is typical of a large class of transitions which hitherto we have not considered.

The specific heat of liquid $_2He^4$ in contact with its vapour, that is, very nearly at constant pressure, varies with temperature in the manner shown in fig. 33. The sharp rise at 2·172° K. corresponds to the transition from He II to He I, and is called, on account of its shape, the λ-point. The most careful measurements have been unable to determine with certainty how high the peak really is, and it is quite probable that an ideal experiment would reveal that it is virtually unbounded. If C_p does in fact tend to infinity at the λ-point it must

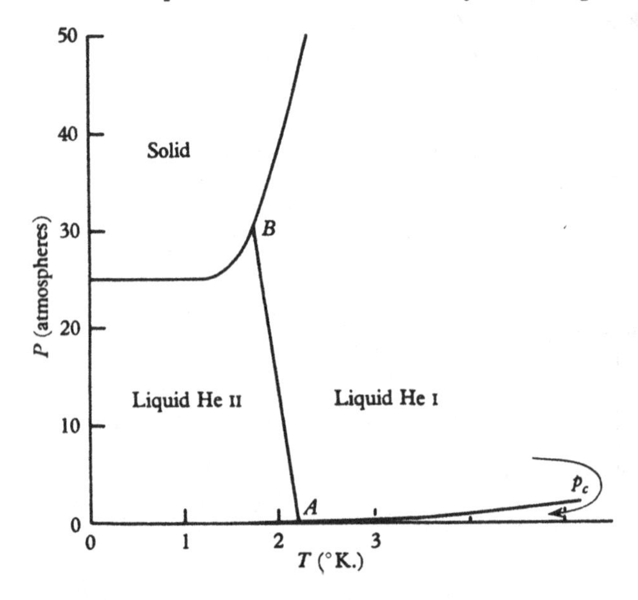

Fig. 32. Phase diagram for $_2He^4$.

do so more slowly than as $(T_\lambda - T)^{-1}$, since the energy change associated with the transition, $\int C_p \mathrm{d}T$, must remain finite. In fact the rise is much slower than this, being represented rather closely by a function of the form $A + B \ln |T_\lambda - T|$ on both sides of the transition, with A taking different values on the two sides;† the specific heat has been determined within a few microdegrees of T_λ. There is no latent heat associated with this transition; if an isolated vessel of He II is supplied with heat at a constant rate, the rate of temperature rise steadily decreases as T_λ is approached, and appears to become momentarily zero at T_λ before increasing suddenly once more, but there is no

† M. J. Buckingham and W. M. Fairbank in *Progress in Low Temperature Physics*, ed. C. J. Gorter (Amsterdam: North Holland, 1961), Vol. III, p. 80.

halting of the temperature for a measurable time at T_λ as would occur if there were a latent heat. The entropy is thus a continuous function of temperature, though there may be a discontinuity in $(\partial S/\partial T)_P$. It is easy to see that the volume is also continuous by applying M. 3 to the liquid. In crossing the λ-line (AB in fig. 32) at constant tempera-

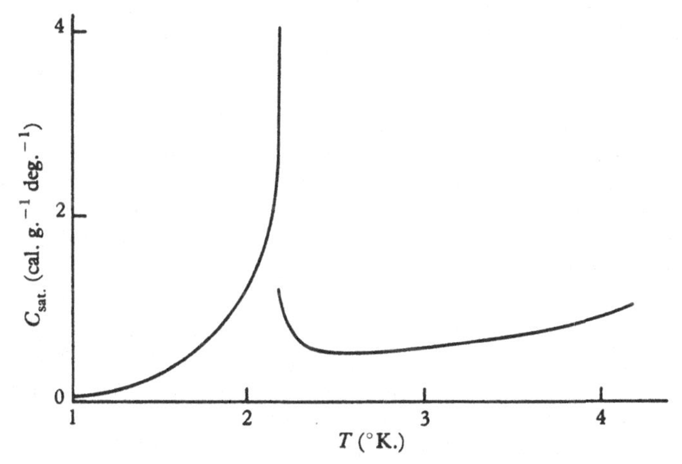

Fig. 33. Specific heat of liquid $_2He^4$ in contact with its vapour.

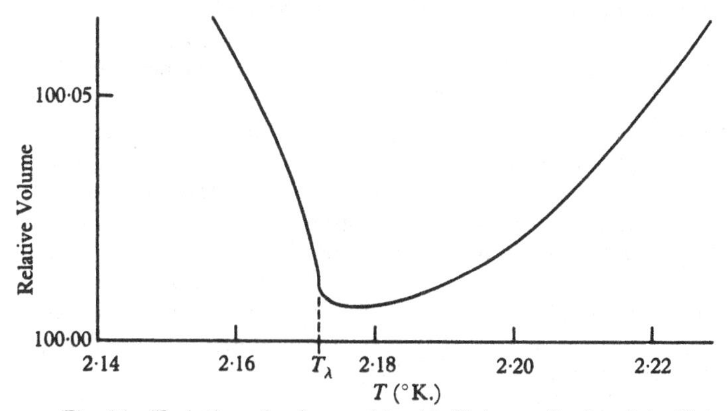

Fig. 34. Variation of volume of liquid $_2He^4$ near the λ-point. Note the vertical tangent at T_λ.

ture the entropy is a continuous function of pressure, although $(\partial S/\partial P)_T$ may become infinite on the line; correspondingly the volume is a continuous function of temperature, although $(\partial V/\partial T)_P$ may become infinite on the line, with the opposite sign to that of $(\partial S/\partial P)_T$. This is illustrated in fig. 34. We have then to deal with a new type of transition which in some respects resembles a smeared-out phase transition—the latent heat is, as it were, absorbed over an interval of temperature instead of at one fixed temperature. It also resembles, and perhaps more closely, the specific heat behaviour at the critical point. Here, however, instead of a single point on the phase diagram at which C_p becomes infinite, we have a line of such points. Clapeyron's equation cannot be applied as it stands to relate the slope of this line to discontinuities of entropy and volume at the transition, since there is no discontinuity of either. We shall consider in the next chapter how to derive thermodynamic relationships analogous to Clapeyron's equation for this type of transition and others.

The Gibbs function for the liquid phase exhibits singular behaviour along the λ-line, but of course in a less striking fashion than the specific heat or entropy. If C_p rises to infinity, the corresponding curve for s shows momentarily a vertical tangent, and the curve for g a point at which the curvature becomes momentarily infinite. The surface $g(P, T)$ is therefore not folded along the λ-line, as it would be if there were a latent heat at the transition, but shows, as it were, an 'incipient fold', the gradient changing rapidly but continuously. The intersection of the g-surface for the liquid with that for the solid, which is quite regular and free from kinks, should in principle reflect the singularity of the liquid in the shape of the liquid-solid equilibrium line. Just as at a triple point, where a regular surface meets a folded surface, there is a sharp change in the gradient of the equilibrium line, so here, where a regular surface meets an almost-folded surface, the equilibrium line exhibits a region of rapidly changing gradient, with one point at which the curvature becomes infinite. The same behaviour should occur at the liquid-vapour equilibrium line. Unfortunately, neither for the liquid-solid nor for the liquid-vapour transition are the experimental data sufficient to show this clearly. Certainly as far as the former is concerned the region of rapidly changing gradient on the melting-curve ought to be readily apparent,† but it would be very hard to recognize on the vapour-pressure curve. The reason for this is easily understood. The volume of unit mass of vapour is a rapidly varying function of temperature, so that the curvature of the vapour-pressure curve is high, as follows from Clapeyron's equation, and the change in entropy of the liquid resulting from the specific heat maximum is small compared with the difference in entropy between the liquid and vapour phases. In consequence, the anomalous variations in curvature resulting from the existence of a λ-transition are comparable with the curvature already present only within a few thousandths of a degree of the λ-point, and give rise to no perceptible deviation in the general trend of the vapour-pressure curve.

† See C. A. Swenson, *Phys. Rev.* **79**, 626 (1950), where the available data are clearly displayed.

In concluding this section the properties of mixtures of the two helium isotopes deserve mention, since they illustrate the third law most strikingly. We have not entered into the subject of chemical thermodynamics, and have thus neglected to introduce the concept of 'entropy of mixing'. We shall therefore not go into details but merely remark that the entropy of a mixture of isotopes is greater than that of the separated isotopes, unless the isotopes can achieve an ordered arrangement, analogous to the ordered state of an alloy shown in fig. 49. The third law demands that the isotopes either order themselves as zero temperature is approached, something that is hard to conceive in a liquid, or else that the two liquids become mutually insoluble. For all that they are so similar it is in fact the latter that occurs. For instance, if a solution containing 3 ${}_2He^3$ atoms to 2 ${}_2He^4$ atoms is cooled, phase separation begins just below 0·9° K. At first a solution richer in ${}_2He^3$ floats on a solution richer in ${}_2He^4$, but with further cooling each phase expels its minority constituent so that as zero temperature is approached the system tends towards the ideal state of zero entropy, pure ${}_2He^3$ floating on pure ${}_2He^4$.

The investigation of liquefied helium isotopes, both pure and mixed, has revealed many new phenomena and contributed notably towards the understanding of the quantum physics of condensed systems, but these are matters beyond our scope and the reader is referred to specialized texts for further study.†

The superconducting phase transition

Certain metals, for example, lead, mercury, tin, aluminium, niobium and tantalum, have the remarkable property of losing all trace of electrical resistance when cooled sufficiently. The disappearance of resistance may take place gradually, over a temperature interval of $\frac{1}{10}$ to 1°, but it is found that if sufficient care is taken to ensure chemical and physical purity, by using single crystals of highly refined metal, the transition usually becomes very sudden. With a good sample of tin, between the point at which the resistance is first observed to diminish and that at which it has entirely vanished, there may lie a temperature interval of only one or two thousandths of a degree. For tin the transition occurs at 3·73° K.; for other elementary metals it varies between 9° K. for niobium and 0·14° K. for iridium. Metal-like compounds have been found which have transition temperatures as high as 18° K., but there seems little doubt that the phenomenon is one which only occurs at temperatures very far below room temperature. As with liquid helium, the chief fascination of the study of superconductors lies in properties which are not strictly

† K. R. Atkins, *Liquid Helium* (Cambridge, 1959).

thermodynamic, and which it is beyond our scope to discuss here. We shall only use superconductivity as an example of a phase transition which involves three independent variables, in this case P, T and $\mathscr{H}$, and because it exhibits under certain circumstances a type of phase transition which is different from any we have encountered hitherto.†

The most important single property of superconductors is not their absence of resistance but their perfect diamagnetism. A magnetic field is unable to enter a superconducting sample, being restrained

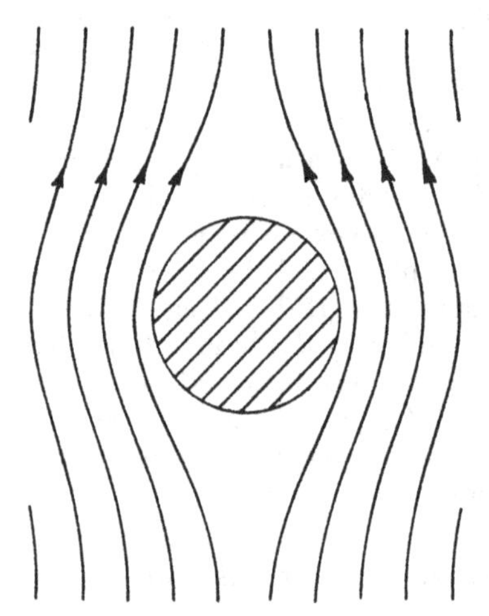

Fig. 35. Magnetic field around superconducting sphere.

from entry by shielding currents which flow in a very thin surface layer. The property of perfect diamagnetism implies that the magnetic field at the surface of a superconductor is everywhere parallel to the surface (a normal magnetic field cannot be prevented from entering and is therefore prohibited by the property of perfect diamagnetism), and if the strength of the field is $\mathscr{H}$ outside and zero inside there must be a surface current density of $\mathscr{H}/4\pi$ (since curl $\mathscr{H} = 4\pi\mathbf{J}$, $\mathbf{J}$ being the volume density of current). A typical example of the magnetic field distribution around a superconductor is shown in fig. 35, where the sample is a sphere. If the mean direction of the field is regarded as defining an axis of the sphere (vertical in fig. 35), the surface currents flow along lines of latitude. Since a magnetic field may be maintained indefinitely by means of a permanent magnet, without any external source of power, it follows that the screening currents are

† For a detailed account of superconductivity see D. Shoenberg, *Superconductivity* (Cambridge University Press, 1952).

entirely non-dissipative, and in this case the property of perfect conductivity is an immediate corollary of the perfect diamagnetism. This is not so obviously true for multiply-connected superconductors, or superconducting wires fed by an external circuit, but it is rather probable that, although no rigorous deduction of perfect conductivity from perfect diamagnetism has been given, it is nevertheless not far from physical reality to regard the latter as the primary property to which the former owes its existence. The present discussion will be confined to simply-connected bodies where this element of doubt is absent. We shall also consider only massive superconductors for which the thickness of the surface layer carrying the screening currents is negligibly small, so that the description of the body as perfectly diamagnetic ($\mathscr{B}=0$) is entirely adequate.

If a long cylindrical superconductor is placed in a magnetic field parallel to its length, the field does not enter until a certain critical field, $\mathscr{H}_c$, is reached. When the critical field is exceeded the metal reverts to its normal, resistive, state and the external field penetrates, so that within the metal $\mathscr{B}=\mathscr{H}$. In reality the susceptibility of the normal metal is not exactly zero, but it is so small that the error involved in putting $\mathscr{B}$ and $\mathscr{H}$ equal is negligible in practice. There is no difficulty in modifying the following arguments to include the normal susceptibility. If the field is slowly reduced from its value greater than $\mathscr{H}_c$ the metal reverts to its perfectly diamagnetic state at the field strength $\mathscr{H}_c$, with expulsion of all the magnetic flux from its interior, a striking phenomenon known, after its discoverers, as the Meissner-Ochsenfeld effect.† Thus the magnetic transition between the superconducting and the normal states is reversible in the thermodynamic sense, and the curve (fig. 36) showing how $\mathscr{H}_c$ depends on temperature

† Or often, with inequitable brevity, simply the Meissner effect. It should be pointed out that this description of the effect is an idealization on two counts. First, it is very commonly observed, particularly with pure metals, that the field must be reduced below $\mathscr{H}_c$ before the superconducting state is re-established. This behaviour is analogous to the supercooling of vapours, and does not occur if a nucleus of superconducting material is formed either in regions of impurity or physical strain, or by artificial means. The phenomenon of supercooling is attributed, as in the vapour-liquid system, to the influence of a high interphase surface energy inhibiting the formation of small nuclei from which the transition may proceed. A second idealization is that the magnetic flux is entirely expelled at the transition. In fact a little is always trapped within the specimen, but in a well-prepared specimen it need not be more than $\frac{1}{10}$% of the flux present before the transition started. This trapped flux is not uniformly distributed over the specimen, but is confined to small channels which do not become superconducting. In the superconducting regions $\mathscr{B}=0$. Neither of these effects invalidates our assumption in what follows that the magnetic behaviour may be idealized as in the text without essentially falsifying the physical picture.

is a transition curve between two phases in the same sense as the term was used in discussing the phase diagram of simple substances.

An extra degree of freedom enters the discussion when it is realized that the transition temperature and the value of $\mathscr{H}_c$ at a given temperature are affected, if only to a small extent, by pressure. If we include the influence of pressure we must regard the state of the metal as determined by three parameters of state, for which T, $\mathscr{H}$ and P are the most convenient choice, and the transition line now becomes extended into a transition surface in $(T, \mathscr{H}, P)$ space. Since the effect of pressure is small (the transition temperature of tin is changed from

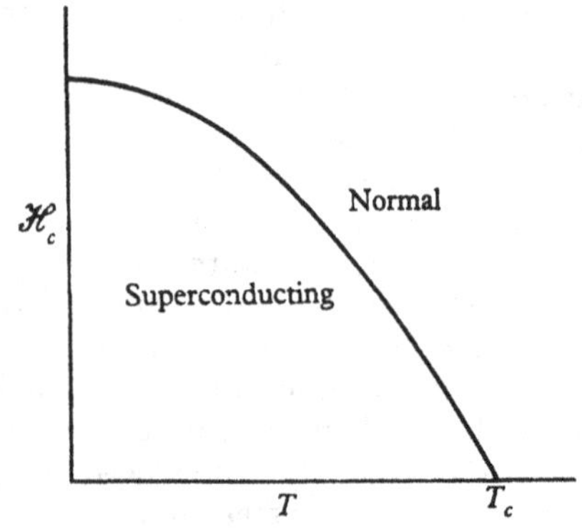

Fig. 36. Variation with temperature of critical magnetic field of a superconductor.

3·73 to 3·63° K. by a pressure of about 1700 atmospheres), sections of the transition surface at different attainable pressures do not differ greatly from the form of fig. 36.

Let us now consider the analogues of Clapeyron's equation for this three-parameter system. We shall confine our attention to a specimen of unit mass in the form of a long cylinder, for which the demagnetizing factor is zero.† It may easily be seen, by extension of the arguments developed in the last section, that the two phases are in equilibrium at such points that their 'magnetic Gibbs functions', $\mathfrak{g}'$, are equal, $\mathfrak{g}'$ being defined by the equation (see (3·13) for the significance of the primes),

$$\mathfrak{g}' \equiv u' - Ts + Pv - \mathscr{H}m, \tag{8·6}$$

where m is the magnetic moment per unit mass. For the superconducting phase $m_s = -v_s\mathscr{H}/4\pi$, since the volume susceptibility of

† See Shoenberg, loc. cit., for a full discussion of shape effects, and a more complete account of the thermodynamic properties of superconductors.

a perfect diamagnetic is $-1/4\pi$, while for the normal phase $m_n=0$ and $\mathrm{g}'_n=g'_n$. From (8·6), since $\mathrm{d}u'=T\,\mathrm{d}s-P\,\mathrm{d}v+\mathscr{H}\,\mathrm{d}m$,

$$\mathrm{d}\mathrm{g}'=-s\,\mathrm{d}T+v\,\mathrm{d}P-m\,\mathrm{d}\mathscr{H}, \tag{8·7}$$

so that

$$\left(\frac{\partial \mathrm{g}'}{\partial T}\right)_{P,\mathscr{H}}=-s,\quad \left(\frac{\partial \mathrm{g}'}{\partial P}\right)_{T,\mathscr{H}}=v\quad\text{and}\quad \left(\frac{\partial \mathrm{g}'}{\partial \mathscr{H}}\right)_{T,P}=-m. \tag{8·8}$$

The analogues of Clapeyron's equation now follow immediately by the same argument as before. By considering a section of the transition surface at a constant pressure we have

$$\left(\frac{\partial \mathscr{H}_c}{\partial T}\right)_P=-\frac{\left(\frac{\partial \mathrm{g}'_n}{\partial T}\right)_{P,\mathscr{H}}-\left(\frac{\partial \mathrm{g}'_s}{\partial T}\right)_{P,\mathscr{H}}}{\left(\frac{\partial \mathrm{g}'_n}{\partial \mathscr{H}}\right)_{P,T}-\left(\frac{\partial \mathrm{g}'_s}{\partial \mathscr{H}}\right)_{P,T}}=-\frac{s_n-s_s}{m_n-m_s}=-\frac{4\pi}{v_s\mathscr{H}_c}(s_n-s_s). \tag{8·9}$$

Similarly
$$\left(\frac{\partial \mathscr{H}_c}{\partial P}\right)_T=\frac{4\pi}{v_s\mathscr{H}_c}(v_n-v_s) \tag{8·10}$$

and
$$\left(\frac{\partial P}{\partial T}\right)_{\mathscr{H}_c}=\frac{s_n-s_s}{v_n-v_s}, \tag{8·11}$$

the last equation being identical with Clapeyron's. The values of s_n, s_s, v_n and v_s to be used in these equations are of course the values taken on the transition surface, although in fact their variation with magnetic field is extremely small. Since, by the magnetic analogue of M.3, $(\partial s/\partial\mathscr{H})_{T,P}=(\partial m/\partial T)_{\mathscr{H},P}$, we see that neither s_n nor s_s are sensibly field-dependent, the former because $m_n=0$ and the latter because $m_s=-v_s\mathscr{H}/4\pi$, independent of temperature, except for a minute effect caused by thermal expansion. And from (8·7) it follows that $(\partial v/\partial\mathscr{H})_{P,T}=-(\partial m/\partial P)_{\mathscr{H},T}$, so that v_n is field-independent and v_s also, apart from an equally minute effect which results from the finite bulk modulus of the superconductor. Thus, except perhaps for the term (v_n-v_s) which is a small difference of large quantities, it is precise enough in practice to use in (8·9), (8·10) and (8·11) the values of s_s and v_s in zero magnetic field, and to regard s_n and v_n as the values which the normal state would possess if it could exist in zero field.

From (8·9) it is seen that the change of entropy at the transition is given by the equation

$$s_n-s_s=-\frac{v_s\mathscr{H}_c}{4\pi}\left(\frac{\partial \mathscr{H}_c}{\partial T}\right)_P. \tag{8·12}$$

Fig. 36 shows that s_n-s_s vanishes at 0° K. in accordance with the third law and also at the critical temperature where $\mathscr{H}_c=0$ and

$(\partial\mathscr{H}_c/\partial T)_P$ is finite. The transition in zero field at T_c is therefore accomplished without any latent heat. This introduces us to a type of transition different from either the simple phase transition or the λ-transition, for here there is a finite discontinuity in specific heat, not infinite as in the λ-transition. This may be seen from (8·12), since

$$c_{P_n} - c_{P_s} = -\frac{v_s T}{8\pi}\left(\frac{\partial^2}{\partial T^2}(\mathscr{H}_c)^2\right)_P. \tag{8·13}$$

Near T_c the critical field is approximately proportional to $T_c - T$, so that the second derivative of $\mathscr{H}_c^2$ is a well-defined finite quantity; since

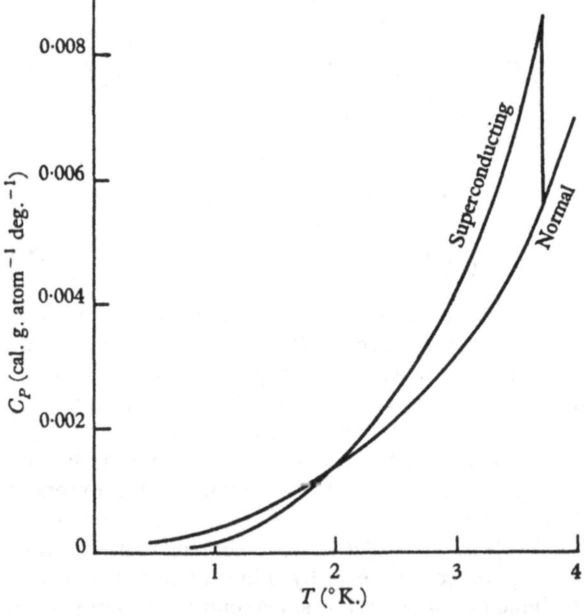

Fig. 37. Specific heat of normal and superconducting tin (W. H. Keesom and P. H. van Laer, *Physica*, 5, 193, 1938).

it is positive, $c_{P_s} > c_{P_n}$ at the transition temperature. The specific heat of tin is shown in fig. 37 for both the normal and superconducting phases. From this experimental curve and (8·13) the variation of $\mathscr{H}_c$ with temperature may be calculated, and is found to agree well with what is measured directly.

Analogous results to (8·13) may be obtained by differentiating (8·10) with respect to pressure and temperature. In this operation only a negligible error will result from taking v_s in the denominator as a constant and equal to v; the results of interest arise from the variations of the small difference $(v_n - v_s)$ caused by changes of pressure and temperature. Thus

$$\frac{1}{8\pi}\left(\frac{\partial^2}{\partial P^2}(\mathscr{H}_c)^2\right)_T = \frac{1}{v}\left\{\left(\frac{\partial v_n}{\partial P}\right)_T - \left(\frac{\partial v_s}{\partial P}\right)_T\right\} = -(k_n - k_s) \qquad (8\cdot14)$$

and
$$\frac{1}{8\pi}\left(\frac{\partial^2}{\partial P\,\partial T}(\mathscr{H}_c)^2\right) = \frac{1}{v}\left\{\left(\frac{\partial v_n}{\partial T}\right)_P - \left(\frac{\partial v_s}{\partial T}\right)_P\right\} = \beta_n - \beta_s, \qquad (8\cdot15)$$

where k is written for the isothermal compressibility, and β for the volume expansion coefficient. The results (8·13)–(8·15) hold at all points on the transition surface. When $\mathscr{H}_c = 0$, i.e. along the transition line in zero field, they can be cast into simpler forms:

$$c_{P_n} - c_{P_s} = -\frac{vT}{4\pi}\left(\frac{\partial \mathscr{H}_c}{\partial T}\right)_P^2, \qquad (8\cdot16)$$

$$k_n - k_s = -\frac{1}{4\pi}\left(\frac{\partial \mathscr{H}_c}{\partial P}\right)_T^2, \qquad (8\cdot17)$$

$$\beta_n - \beta_s = \frac{1}{4\pi}\left(\frac{\partial \mathscr{H}_c}{\partial P}\right)_T\left(\frac{\partial \mathscr{H}_c}{\partial T}\right)_P, \qquad (8\cdot18)$$

from which it follows, by combining these equations in pairs, that

$$\left(\frac{\partial T_c}{\partial P}\right)_{\mathscr{H}_c=0} = vT_c\frac{\beta_n - \beta_s}{c_{P_n} - c_{P_s}} = \frac{k_n - k_s}{\beta_n - \beta_s}, \qquad (8\cdot19)$$

since
$$\left(\frac{\partial \mathscr{H}_c}{\partial P}\right)_T \Big/ \left(\frac{\partial \mathscr{H}_c}{\partial T}\right)_P = -\left(\frac{\partial T_c}{\partial P}\right)_{\mathscr{H}_c}.$$

The experimental verification of these results is not easy, for although the change in specific heat is measurable with considerable accuracy, the expansion coefficients in both states are very small indeed, while on account of the small value of $\partial T_c/\partial P$ (itself not easily measured) the compressibility changes by only a few parts in a million. Such data as have been obtained, however, agree satisfactorily with (8·19). These equations, which we have derived by considering the special case of a superconductor, are in fact the analogues to Clapeyron's equation applicable, as we shall see in the next chapter, to any transition in which there is no latent heat and a finite discontinuity in C_P.

CHAPTER 9

HIGHER-ORDER TRANSITIONS

Classification of transitions

In the last chapter we noted three distinct types of thermal behaviour occurring along lines separating different phases or modifications of a substance, the normal transition with latent heat, the λ-transition without latent heat but a very high (perhaps infinite) peak of specific heat, and the so-called second-order transition in which there is no latent heat and a finite discontinuity in specific heat. The first and last are members of a classification introduced by Ehrenfest, in which the 'order' of a transition is determined by the lowest order of differential coefficient of the Gibbs function which shows a discontinuity on the transition line. Thus in a phase transition which involves latent heat, g is continuous across the line, but its derivatives $(\partial g/\partial T)_P$ and $(\partial g/\partial P)_T$, $-s$ and v respectively, are discontinuous; such a transition is said to be of the first order. In the transition of a superconductor in zero magnetic field there is no latent heat and no volume change, so that the first derivatives of g are continuous, but the second derivatives, representing specific heat, expansion coefficient and compressibility are discontinuous, so that this is a transition of the second order. The classification may be extended indefinitely, though as the order of the transition increases it becomes less and less clear that it is appropriate to think of the process as a change from one phase to another since the discontinuity in properties becomes progressively less significant. For instance, in a third-order transition the specific heat is a continuous function of temperature, only its gradient showing a sharp break, while in a fourth-order transition the C_P-T curve is distinguished merely by possessing a discontinuity of curvature. Thus in practice it is only the first- and second-order transitions which usually arouse interest, and we shall mainly confine our discussion to these.

Now this classification of Ehrenfest's, while having served a valuable purpose in pointing to a distinction between different types of transition, is of only limited application, since true second-order transitions are exceedingly unusual. It is probably true to say that of physically interesting systems (that is, excluding *ad hoc* models, of which we shall discuss one in detail) there is only one class, the superconducting transition, which bears any resemblance to an ideal second-order transition. On the other hand, there are many transitions known to occur in widely different varieties of substance which do not comfortably conform to Ehrenfest's scheme. The transition in liquid helium is an example, and others will be mentioned later. A few conceivable specific heat curves are drawn diagrammatically in fig. 38,

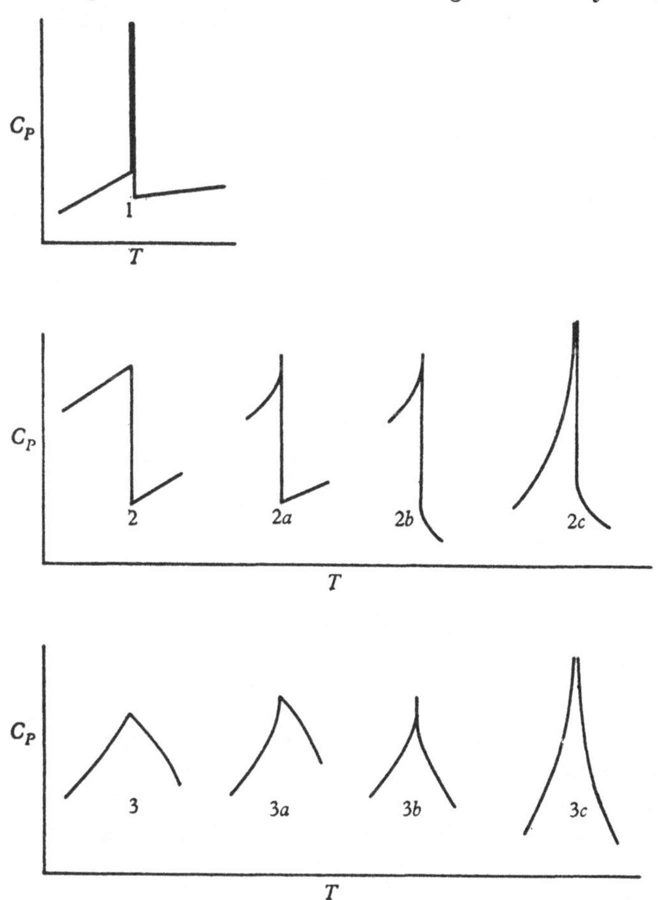

Fig. 38. Classification of transitions.

and arranged so as to present the variants of Ehrenfest's classification. Type 1 is the simple first-order transition, with latent heat. Types 2, 2 *a*, *b* and *c* show no latent heat but a discontinuity in C_P; in 2 *a* $\partial C_P/\partial T$ tends to infinity as the transition is approached from one side, in 2 *b* it tends to infinity from both sides; in 2 *c* the discontinuity in C_P is infinite. The third-order transitions are analogous variants of the standard Ehrenfest third-order transition (type 3).

Not all these types have been observed in practice, but there appears to be no thermodynamical reason why they should not occur. A few examples of those which have been observed are given below,† together with some idealized theoretical models which have been calculated exactly and found to exhibit interesting transitions. The latter are placed in square brackets.

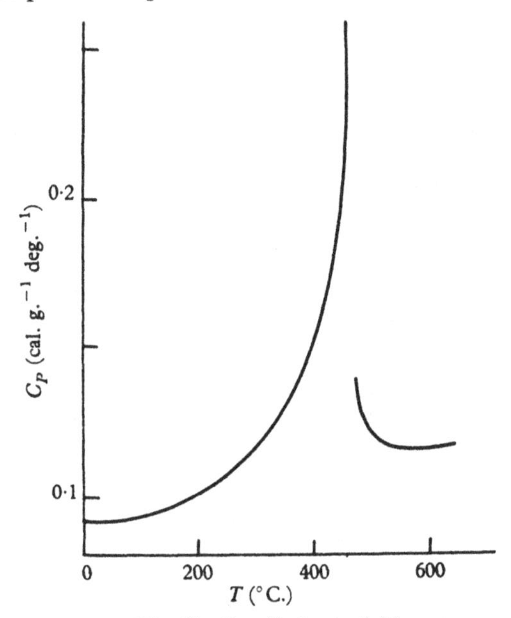

Fig. 39. Specific heat of β-brass.

(1) First-order transitions: Solid-liquid-vapour transitions. Many allotropic transitions in solids, e.g. grey to white tin.

(2) Second-order transitions: Superconducting transition in zero field (fig. 37). [Weiss model of ferromagnetism,‡ Bragg-Williams model of the order-disorder transformation in β-brass.§]

(2c) λ-transitions: Order-disorder transformation in β-brass‖ (fig. 39), ammonium salts,¶ crystalline quartz‖ (fig. 40), solid hydrogen†† (fig. 41) and many other solids.

† For an account of the phenomena of ferromagnetism, antiferromagnetism and order-disorder transitions, consult C. Kittel, *Introduction to Solid State Physics* (Wiley, 1953).

‡ F. Seitz, *Modern Theory of Solids* (McGraw-Hill, 1940), p. 608.

§ *Ibid.* p. 507.

‖ H. Moser, *Phys. Z.* **37**, 737 (1936).

¶ F. E. Simon, C. von Simson and M. Ruhemann, *Z. Phys. Chem.* A, **129**, 339 (1927).

†† R. W. Hill and B. W. A. Ricketson, *Phil. Mag.* **45**, 277 (1954).

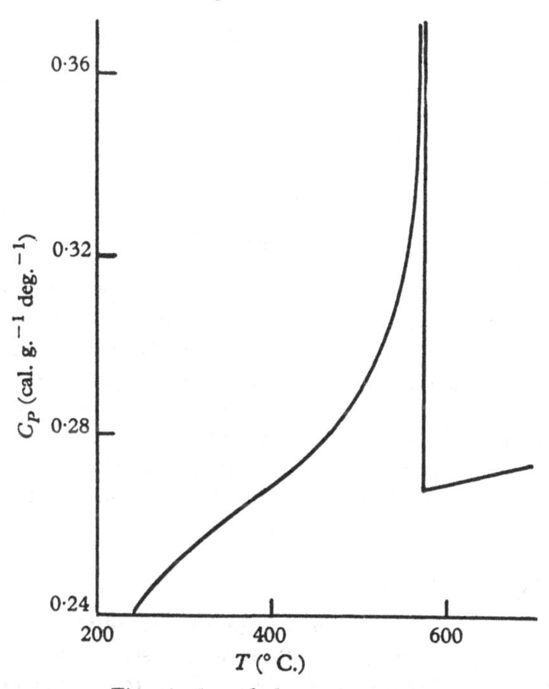

Fig. 40. Specific heat of crystalline quartz.

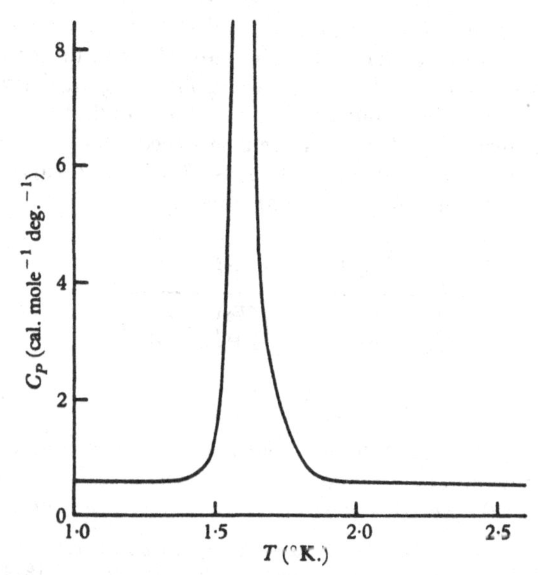

Fig. 41. Specific heat of solid hydrogen containing 74 % orthohydrogen, 26 % parahydrogen.

(3) Third-order transitions: The Curie points of many ferromagnetics† (fig. 42).

(3c) Symmetrical λ-transitions: The antiferromagnetic transition in $MnBr_2$‡ (fig. 43). Liquid $_2He^4$ (fig. 33). [Two-dimensional Ising model of the order-disorder transformation.§]

One might also include in category (3c) the critical point of liquid-vapour systems, as discussed in the last chapter. This is rather different from the rest, however, as the infinity in C_P occurs only at one point, and not along a transition line.

It is well to remember that the distinctions made between different types of transition in this catalogue may prove to have been too nice. It is very easy for the character of a transition to be falsified through imperfections in the samples or experimental inaccuracy, and it is not unlikely that almost every transition which is not of first order would be found to be of type 3c under ideal conditions. The only exception to this statement is the superconducting transition, which obstinately persists in exhibiting a finite discontinuity in C_P in the most refined experiments.

Analogues of Clapeyron's equation

For all transitions except those of the first order, not only the Gibbs function but the entropy is continuous across the transition line, since there is no latent heat. It follows at once from Clapeyron's equation, since in general the slope of the transition line, $\mathrm{d}P/\mathrm{d}T$, is finite and non-zero, that the volume also is continuous, and Clapeyron's equation degenerates to the form $\mathrm{d}P/\mathrm{d}T=0/0$. But now the continuity of s and v enables arguments similar to those applied to g in deriving Clapeyron's equation to be applied to s and v. We shall first consider a second-order transition of type 2 (Ehrenfest's second-order transition). For this s and v are continuous, but $(\partial s/\partial T)_P$ is not on account of the jump in C_P. We therefore write immediately,

$$\frac{\mathrm{d}P}{\mathrm{d}T}=-\frac{\left(\frac{\partial s_2}{\partial T}\right)_P-\left(\frac{\partial s_1}{\partial T}\right)_P}{\left(\frac{\partial s_2}{\partial P}\right)_T-\left(\frac{\partial s_1}{\partial P}\right)_T}=-\frac{\left(\frac{\partial v_2}{\partial T}\right)_P-\left(\frac{\partial v_1}{\partial T}\right)_P}{\left(\frac{\partial v_2}{\partial P}\right)_T-\left(\frac{\partial v_1}{\partial P}\right)_T},$$

or, from M. 3,

$$\frac{\mathrm{d}P}{\mathrm{d}T}=\frac{1}{vT}\frac{c_{P_2}-c_{P_1}}{\beta_2-\beta_1}=\frac{\beta_2-\beta_1}{k_2-k_1}, \tag{9·1}$$

in which, as usual, β is the volume expansion coefficient and k the isothermal compressibility. Equations (9·1), which are of course identical with (8·19), are Ehrenfest's equations for a second-order

† H. Moser, loc. cit.; J. B. Austin, *Industr. Engng Chem.* **24**, 1225 (1932).
‡ W. P. Hadley, private comm. § L. Onsager, *Phys. Rev.* **65**, 117 (1944).

transition. We discussed in the last chapter the degree of success with which they have been verified for the superconducting transition, the only type of transition to which they may strictly be applied.

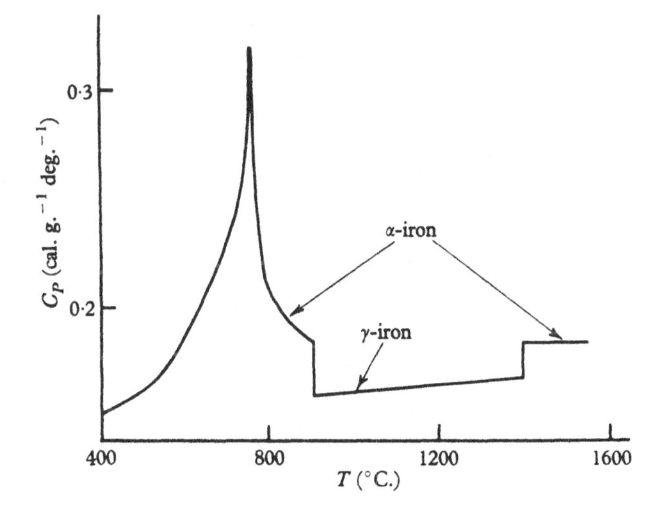

Fig. 42. Specific heat of iron.

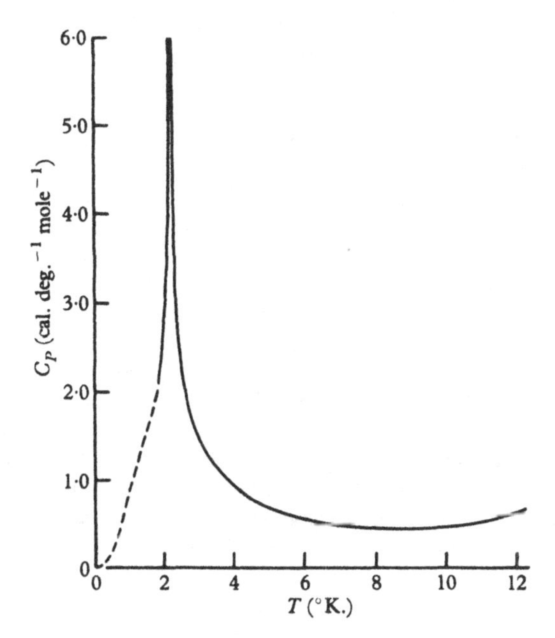

Fig. 43. Specific heat of anhydrous $MnBr_2$.

For Ehrenfest's third-order transition (type 3), not only g, s and v are continuous, but also c_P, β and k, while the derivatives of the latter are discontinuous. There are in consequence three analogues of Clapeyron's equation for a third-order transition, which the reader will easily verify to have the form

$$\frac{\mathrm{d}P}{\mathrm{d}T}=\frac{1}{vT}\frac{\left(\frac{\partial c_{P_2}}{\partial T}\right)_P-\left(\frac{\partial c_{P_1}}{\partial T}\right)_P}{\left(\frac{\partial \beta_2}{\partial T}\right)_P-\left(\frac{\partial \beta_1}{\partial T}\right)_P}=\frac{\left(\frac{\partial \beta_2}{\partial T}\right)_P-\left(\frac{\partial \beta_1}{\partial T}\right)_P}{\left(\frac{\partial k_2}{\partial T}\right)_P-\left(\frac{\partial k_1}{\partial T}\right)_P}=\frac{\left(\frac{\partial \beta_2}{\partial P}\right)_T-\left(\frac{\partial \beta_1}{\partial P}\right)_T}{\left(\frac{\partial k_2}{\partial P}\right)_T-\left(\frac{\partial k_1}{\partial P}\right)_T}. \tag{9·2}$$

Relations between higher derivatives similarly exist for transitions of higher order. No experimental verification has been attempted for any transitions of higher than second order.

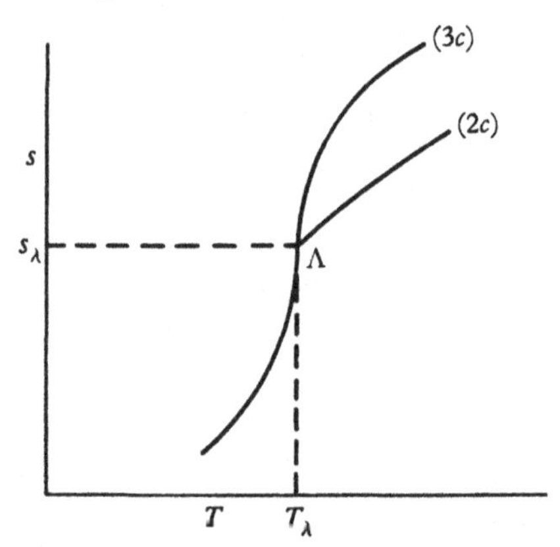

Fig. 44. Variation of entropy with temperature at a λ-point.

Since in transitions of the types (2*a*) and (2*b*) the discontinuities in c_P are finite, even though $\partial c_P/\partial T$ may become infinite at the transition, Ehrenfest's equations for the second-order transition are applicable to them. On the other hand, the infinities in c_P for types (2*c*) and (3*c*) and the infinities in $\partial c_P/\partial T$ for types (3*a*) and (3*b*) convert equations like (9·1) and (9·2) into indeterminacies of the form ∞/∞. Although this circumstance rules out Ehrenfest's method of approach, it enables an alternative approach to be made, which may be illustrated by consideration of types (2*c*) and (3*c*). Here the specific heat rises to infinity, so that as the transition is approached from below $(\partial s/\partial T)_P$ and $(\partial^2 s/\partial T^2)_P$ tend to infinity, and the variation of entropy with temperature, at constant pressure, takes the form shown diagrammatically in fig. 44. At the transition temperature T_λ, the curve reaches a point of inflexion, Λ, with vertical tangent, and either continues without discontinuity of gradient, for a transition of type ($3c$), or breaks sharply at the inflexion, for a transition of type $2c$. The value, s_λ, of the entropy at T_λ will normally be a smoothly varying function of pressure, so that the surface $s(T, P)$ will have a regular fold along the transition line, and at any point near this fold (on either side for ($3c$), but only on the left-hand side for ($2c$)) there will be a very great second derivative in a plane normal to the transition line, and only a much smaller one in a plane tangential to the transition line. We may therefore hope to approximate to the shape of the entropy surface by treating it, over a short range of pressure, and in the vicinity of T_λ, as a cylindrical surface:

$$s = s_\lambda(P) + f(P - \alpha T). \tag{9·3}$$

In this approximation it is taken to be adequate to treat the line of inflexion points Λ as straight; this is equivalent to the assumption that the second derivative of this line is negligible in comparison with the second derivative of the function f which describes the shape of the curve in fig. 44. Clearly α is equal to $(\mathrm{d}P/\mathrm{d}T)_\lambda$, the slope of the transition line. Then, since α is taken as constant and $s_\lambda(P)$ to contain no quadratic or higher terms, (9·3) yields the results

$$\left(\frac{\partial^2 s}{\partial T^2}\right)_P = \alpha^2 f'', \quad \left(\frac{\partial^2 s}{\partial T \partial P}\right) = -\alpha f'' \quad \text{and} \quad \left(\frac{\partial^2 s}{\partial P^2}\right)_T = f'', \tag{9·4}$$

in which f'' is the second derivative of f with respect to its argument. Hence

$$\left(\frac{\mathrm{d}P}{\mathrm{d}T}\right)_\lambda = \alpha = -\left(\frac{\partial^2 s}{\partial T^2}\right)_P \Big/ \left(\frac{\partial^2 s}{\partial T \partial P}\right) = -\left(\frac{\partial^2 s}{\partial T \partial P}\right) \Big/ \left(\frac{\partial^2 s}{\partial P^2}\right)_T,$$

or, by use of M. 3,

$$\frac{\partial}{\partial T}\left(\frac{\partial s}{\partial T}\right)_P = \alpha \frac{\partial}{\partial T}\left(\frac{\partial v}{\partial T}\right)_P \quad \text{and} \quad \frac{\partial}{\partial P}\left(\frac{\partial s}{\partial T}\right)_P = \alpha \frac{\partial}{\partial P}\left(\frac{\partial v}{\partial T}\right)_P. \tag{9·5}$$

The results (9·5) imply that in the vicinity of the transition line, $(\partial s/\partial T)_P$ is a linear function of $(\partial v/\partial T)_P$, or that

$$c_P = \alpha v T_\lambda \beta + \text{const.} \tag{9·6}$$

It follows from (9·6) that as c_P tends to infinity, so does β, and therefore that the surface representing $v(T, P)$ also has a fold like that for $s(T, P)$. The same argument may then be applied to $v(T, P)$, to yield the equation

$$\beta = \alpha k + \text{const.} \tag{9·7}$$

Equations (9·6) and (9·7) are based upon a cylindrical approximation to the form of the entropy and volume surfaces, and may not apply at any great distance from T_λ; they should, however, become increasingly more exact as the λ-point is approached.

It will be noted that there is a marked similarity between these equations and Ehrenfest's equations (9·1), and indeed Ehrenfest's equations are readily derived from (9·6) and (9·7). At a transition of type 2 the entropy surface has a sharp fold where the gradient changes abruptly. If this fold is smoothed out into a cylinder of high curvature,

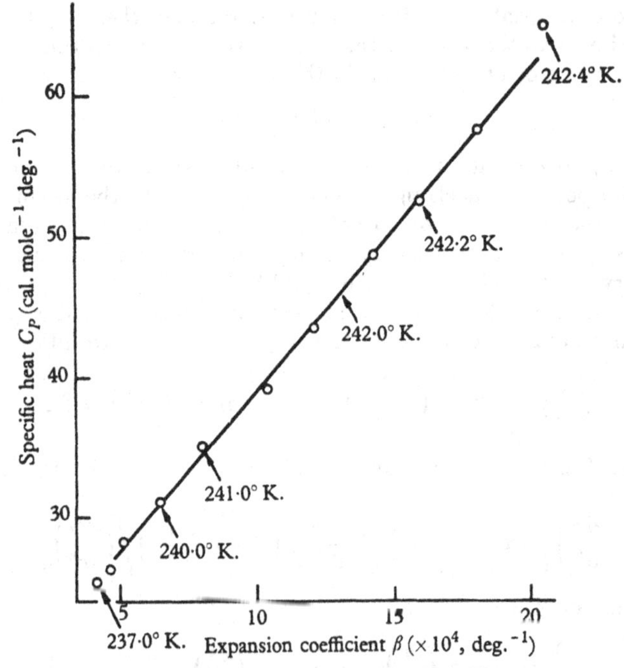

Fig. 45. C_P v. β for NH_4Cl in vicinity of λ-point. The temperatures involved are shown along the line.

(9·6) holds on this cylinder, so that between one side of the cylinder and the other $\Delta C_P = \alpha v T_\lambda \, \Delta\beta$; when the radius of curvature of the cylinder is allowed to tend to zero, this result becomes identical with one of Ehrenfest's equations. The other is similarly derived from (9·7).

There is very little experimental material available to provide a test of these equations, but the λ-transition in ammonium chloride has been sufficiently studied to enable (9·6) to be applied to it. In fig. 45 values of C_P and β measured at the same temperature are plotted against one another, and it is clear that the linear relation predicted from (9·6) holds over an interval of several degrees below T_λ (the data above T_λ are insufficiently reliable to be useful). From the slope of the line α may be found, and it is predicted that the λ-temperature should be raised by 1° by applying a pressure of 113 atmospheres. Fig. 46 shows the only measurements of T_λ as a function of pressure. In view of the scanty data the agreement of predicted and measured slopes is satisfactory. Much more information is available concerning the λ-transition in liquid helium, but it turns out that the slope deduced for α when C_p is plotted against β, or β against k, may be incorrect to the extent of a factor of nearly two if the temperature scales used in

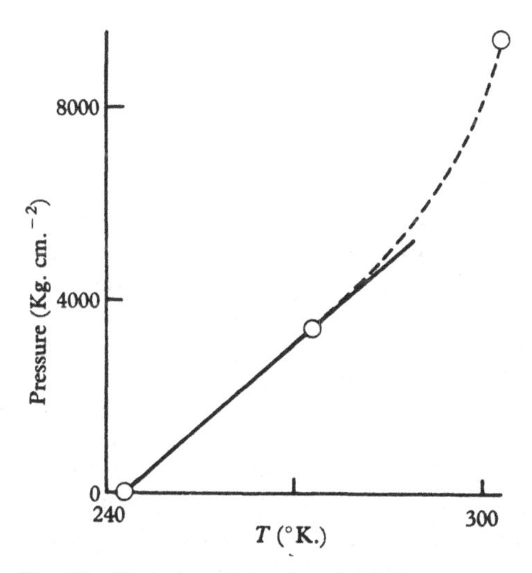

Fig. 46. Variation of λ-point of NH_4Cl with pressure. The line is calculated from (9·6).

the determination of these quantities (by different groups of workers) differ from one another by as little as 0·001° K. As an error of this magnitude is only too likely in practice, we cannot regard liquid helium as providing a satisfactory test. However, the validity of (9·6) and (9·7) is not open to serious doubt in the vicinity of a λ-point, and it is possible that their chief application lies in testing critically the experimental data for a transition such as that of helium, and in eliminating relative errors in the temperature scales of different workers.†

We have seen earlier that, although c_p may become infinite, c_V must

† For further discussion, see the article by Buckingham and Fairbank referred to on p. 126 (footnote). For the application of (9·6) and (9·7) to the α–β transition in quartz, see A. J. Hughes and A. W. Lawson, *J. Chem. Phys.* **36**, 2098 (1962).

remain finite (except under the most unusual circumstances), and this analysis of the λ-transition provides a convenient illustration. Writing (9·6) and (9·7) in the form

$$c_P - c_0 = \alpha v T_\lambda \beta,$$
$$\beta - \beta_0 = \alpha k,$$

and making use of (6·10), we find that near the λ-point

$$c_P - c_V = (c_P - c_0)\left(1 - \frac{\beta_0}{\beta}\right)^{-1}$$
$$= (c_P - c_0)\left(1 + \frac{\beta_0}{\beta} + \frac{\beta_0^2}{\beta^2} + \dots\right),$$

or

$$c_V = (c_0 - \alpha v T_\lambda \beta_0) - \alpha v T_\lambda \beta_0 \left(\frac{\beta_0}{\beta} + \dots\right). \tag{9·8}$$

Now since the area under the curve of c_P must be finite, c_P, and hence β, must go to infinity less rapidly than $(T_\lambda - T)^{-1}$, so that $1/\beta$ must go to zero with a vertical tangent at T_λ. This means that as the λ-point is approached c_V rises (if α is positive) with a vertical tangent to the finite value given by the first bracket in (9·8). The contrast between c_P and c_V is thus very marked, and indeed in ammonium chloride an analysis of the data leads to no evidence of any significant change in c_V as T_λ is approached. This result has importance in connexion with attempts to analyse in detail the mechanisms responsible for λ-points. In most calculations by statistical methods it is much easier to handle a system maintained at constant volume than one at constant pressure. But it must always be remembered that in the former the symptoms of a λ-transition may be much less pronounced than in the latter, and it may even be possible to lose the transition altogether by employing approximate means of calculation on a constant-volume system.

Critique of the theory of higher-order transitions

After Ehrenfest's thermodynamical treatment of second-order transitions was published it became a target for several critical attacks which, taken at face value, appeared to be not without substance. Following these attacks the literature on the subject has become rather unnecessarily confused, and we shall attempt by the following discussion to remove some of the resulting obscurity. A complete discussion can be made more easily with the aid of statistical mechanics, but we shall avoid this elaboration by analysing artificial models which may be regarded as analogues of the real physical systems of interest.

First let us look at the criticisms of the theory, of which the two of importance are essentially concerned with pointing out a contrast between transitions of the first and second orders. Fig. 47 shows how the Gibbs function varies with temperature for the two types of transition. For the first-order transition the curves for g in the two phases (at a given pressure) cut each other at the equilibrium temperature, so that phase 1 is stable below and phase 2 above the equilibrium temperature. In the second-order transition, since there is no entropy difference between the two phases at the equilibrium temperature, but a difference in specific heat, the two curves osculate, and the difference in curvature at the point of contact ensures that there is no cross-over (in a third-order transition there would be a three-point

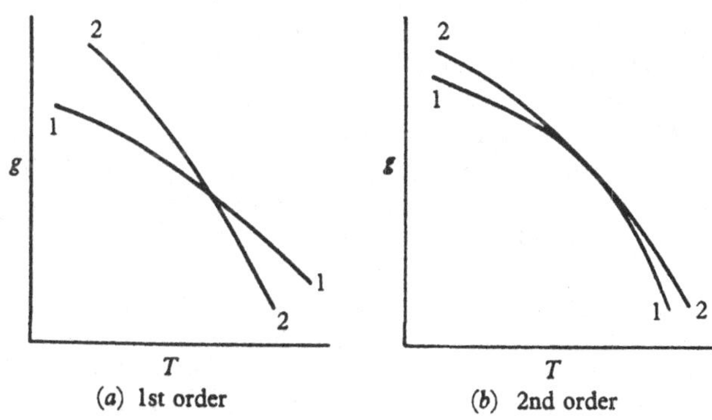

Fig. 47. Suggested variation of the Gibbs function in first- and second-order transitions.

contact at the equilibrium temperature and the curves would cross). It may now be pointed out in criticism of the conception of a second-order transition that

(1) whereas two lines in general cross each other at some point, the chance of their meeting so as to osculate exactly is so small as never to be observed in the world of physical phenomena;

(2) if the two lines osculate without crossing, the line corresponding to phase 1 remains below that for phase 2 both above and below the equilibrium temperature, and phase 1 is therefore stable at all temperatures.

On both these counts the critics of Ehrenfest's theory have claimed that the second-order transitions cannot occur in nature, in spite of the fact that the superconducting transition appears to provide convincing refutation of their view.

It is now convenient to analyse in some detail a simple physical system which simulates a second-order transition, in order to see how the criticisms can be explained away. This model of a second-order transition was suggested by Gorter, who did not give a detailed analysis, no doubt regarding its behaviour as obvious to the instructed imagination. As however even this model has been denounced as imperfect it will be as well, and in addition it will provide an instructive example of thermodynamical reasoning, to show how it leads to Ehrenfest's equations. Gorter's model is depicted in fig. 48. A vessel contains a small amount of liquid, the remaining space being filled with its vapour; the walls of the vessel have negligible thermal capacity and are not quite inextensible, so that the volume of the vessel is altered by a variation of the difference, Π, between the external pressure P and internal pressure P_i. The vapour is assumed to behave as a perfect gas, having C_P and C_V independent of pressure and temperature. As the vessel is warmed, P being constant, liquid evaporates into the vapour phase so as to keep P_i equal to the vapour pressure, and the effective thermal capacity of the vessel includes not only the thermal capacities of liquid and vapour but also the latent heat of evaporation. At a certain temperature, T_0, the last drop of liquid evaporates (the amount of liquid is adjusted so that this occurs below the critical point), and thereafter there is no latent heat contribution to the effective thermal capacity, which thus shows a sharp drop, as in a second-order transition. At the same time the internal pressure P_i, which up to T_0 had been the vapour pressure, now begins to increase in proportion to T, following the perfect gas law. There is thus a discontinuity in $(\partial\Pi/\partial T)_P$, and hence in $(\partial V/\partial T)_P$, just as in a second-order transition. It will also be seen that the compressibility shows a discontinuity at T_0.

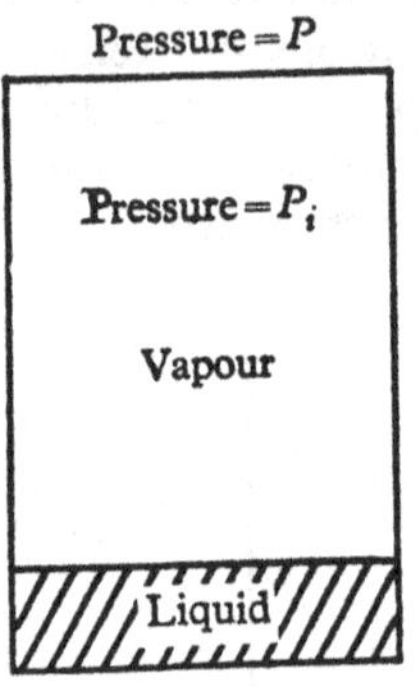

Fig. 48. Gorter's model of a second-order transition.

We now analyse the model in detail (we shall leave the reader to verify some of the intermediate steps of the calculation). When $P=0$, let T_0 be the 'transition' temperature at which the last drop of liquid disappears, and correspondingly let the internal pressure be P_0 and the volume of the vessel V_0. Let the elasticity of the vessel be such that $(\mathrm{d}V/\mathrm{d}\Pi) = -a$. First we calculate how the transition temperature depends on P. When P is increased from zero to δP, let the transition temperature change from T_0 to $T_0+\delta T$, and the internal pressure at the transition temperature from P_0 to $P_0+\delta P_i$. Then at the transition temperature P_i is equal to the vapour pressure, and from Clapeyron's equation,

$$\delta P_i = A\delta T,$$

where
$$A = \frac{l}{T_0(v_v - v_l)}.$$

Corresponding to the changes δP and δP_i, V changes by δV, where

$$\delta V = a(\delta P_i - \delta P) = a(A\delta T - \delta P).$$

Now at the transition all the material is in the vapour phase, to which the perfect gas law applies, so that

$$P_0 \delta V + V_0 \delta P_i = R\,\delta T,$$

or
$$\delta T = \frac{aP_0}{A(V_0 + aP_0) - R}\,\delta P.$$

Thus the slope dP/dT of the transition line is given by the expression

$$\alpha \equiv dP/dT = \frac{A(V_0 + aP_0) - R}{aP_0}. \tag{9·9}$$

To find the expansion coefficient β, we note that expansion is caused (P being constant) by the increase of P_i with temperature. Below T_0, by Clapeyron's equation dP_i/dT is simply A; above T_0, by the gas law, dP_i/dT is $R/(V_0 + aP_0)$. Hence the discontinuity in β is given by the expression

$$\Delta\beta = \frac{a}{V_0}\,\frac{R - A(V_0 + aP_0)}{V_0 + aP_0}. \tag{9·10}$$

A similar analysis of the variation of V caused by changing P leads to the discontinuity in isothermal compressibility,

$$\Delta k = -\frac{a^2 P_0}{V_0(V_0 + aP_0)}. \tag{9·11}$$

From (9·9), (9·10) and (9·11) it will be seen that one of Ehrenfest's equations is satisfied, $\Delta\beta/\Delta k = \alpha$.

To verify the other of Ehrenfest's equations we must calculate the discontinuity in specific heat, ΔC_P, and for this purpose it is convenient to write down expressions for the entropy of the material in the vessel. Let there be unit mass of material altogether, and let s_l and s_v be the entropy if all the material were in the liquid and vapour phase respectively at temperature T_0 and pressure P_0. Then $T_0(s_v - s_l) = l$. At a temperature $T_0 - \delta T$ the entropy of the liquid at its saturation vapour pressure would be $s_l - c_{P_l}\delta T/T_0$, while that of the vapour would be

$s_v-(c_{P_v}-AV_0)\,\delta T/T_0$.† If then at $T_0-\delta T$ an amount δq of liquid is present, the entropy of the system takes the form

$$S=s_v-\frac{l}{T_0}\,\delta q-\frac{c_{P_v}-AV_0}{T_0}\,\delta T.$$

Since it follows from the gas law and Clapeyron's equation that

$$\delta q=AV_0\frac{A(aP_0+V_0)-R}{Rl}\,\delta T,$$

therefore

$$S=s_v-\left\{\frac{A^2V_0}{R}(aP_0+V_0)-2AV_0+c_{P_v}\right\}\frac{\delta T}{T_0},$$

and the effective thermal capacity just below T_0,

$$C_{P1}=\frac{A^2V_0}{R}(aP_0+V_0)-2AV_0+c_{P_v}.$$

The thermal capacity just above T_0 is easily seen to take the form

$$C_{P2}=c_{P_v}-\frac{RV_0}{aP_0+V_0}.$$

Hence $$\Delta C_P\equiv C_{P1}-C_{P2}=\frac{T_0}{P_0}\frac{\{A(aP_0+V_0)-R\}^2}{aP_0+V_0}. \qquad (9\cdot12)$$

From (9·9), (9·10) and (9·12) it follows that $\Delta C_P/\Delta\beta=V_0T_0\alpha$, in agreement with Ehrenfest's equation.

This model serves to show that there is no intrinsic violation of thermodynamic principles involved in the existence of a second-order transition, and we may now examine the criticisms mentioned above in the light of the model. It is immediately clear that they are based on an erroneous analogy with first-order transitions. To take the second criticism first, at temperatures below that at which the lines 1 and 2 (fig. 47*b*) meet, the stable phase 1 in Gorter's model is that in which the vessel contains both liquid and vapour, while phase 2 corresponds to supercooled vapour only. As the transition temperature is approached, the amount of liquid steadily diminishes until at the point of contact of the lines it finally vanishes. The continuation of line 1 above this

† $$\mathrm{d}s_v=\left(\frac{\partial s_v}{\partial T}\right)_P\mathrm{d}T+\left(\frac{\partial s_v}{\partial P}\right)_T\mathrm{d}P$$

$$=\frac{c_{P_v}}{T}\mathrm{d}T-\left(\frac{\partial v_v}{\partial T}\right)_P\mathrm{d}P \quad \text{from M.3}$$

$$=\frac{c_{P_v}-Av}{T}\,\mathrm{d}T \quad \text{from the gas law and Clapeyron's equation.}$$

temperature is physically absurd—it could only mean a state in which the vessel contains a negative amount of liquid. Thus the true diagram should rather consist of a single line at temperatures above the transition, which breaks into two branches below the transition temperature. By the same token the first criticism is also disposed of. The two lines do not correspond to totally different phases, such as liquid and vapour, which can be considered independently of one another so that the phase transition results from an almost fortuitous meeting of the lines; rather, the two phases become more and more similar in constitution as the transition temperature is approached, and at that temperature they are actually identical.

We may see from this analysis what is the characteristic difference between first-order transitions and transitions of the second and higher orders. The two phases which are in equilibrium at a first-order transition are different in their physical constitution, greatly different when they are solid and vapour, not so greatly when they are solid and liquid, but always sufficiently different that they possess different energies, entropies and volumes. In a second-order transition, on the other hand, the two phases (and at this stage we may begin to doubt the wisdom of using this terminology to describe the two states on either side of the transition line) are identical in constitution, energy, entropy and volume. Where they differ is in the rates of variation with respect to temperature, pressure, etc., of these primary thermodynamical parameters. In fact whereas a first-order transition marks the point at which a major change in properties occurs, a second-order transition only marks the point at which a change begins to occur. Thus in Gorter's model the jump in specific heat (as the temperature is decreased) occurs when the liquid phase begins to condense. Similarly the transition in β-brass, according to the simplified description of Bragg and Williams, marks the onset of an ordering process, not the establishment of an ordered state. Above the transition temperature copper and zinc atoms, present in equal numbers, are arranged at random on a body-centred cubic lattice; at very low temperatures the two types of atom take up a completely ordered arrangement on alternate lattice sites, as shown in fig. 49. At the transition temperature itself the arrangement is still quite random, in the sense that if we count the number of copper and zinc atoms on one set of sites (those denoted by open circles for example) we shall find them to be equal. But as the temperature is lowered the tendency begins to show for the copper atoms to favour one set of sites and the zinc atoms the other, and this tendency becomes more pronounced as the temperature falls until at 0° K. there is complete order. It should be emphasized that this is a simplified theoretical model which leads, when worked out in detail, to a second-order transition of the Ehrenfest type, while the reality conforms more nearly to a transition of type 2*c*. This does not, however, detract from its value as an illustration of the way in which a second-order transition marks the beginning, not the achievement, of a change in character of a physical system.

We have seen how the critics of higher-order transitions were misled by a false analogy with first-order transitions, so that they implicitly assumed the existence of two g-surfaces, one for each 'phase' and each capable, as it were, of existing independently of the other. The justification for this assumption in the theory of first-order transitions, as was discussed in Chapter 8, essentially lies in the possibility of establishing superheated and supercooled states, and thus of demonstrating

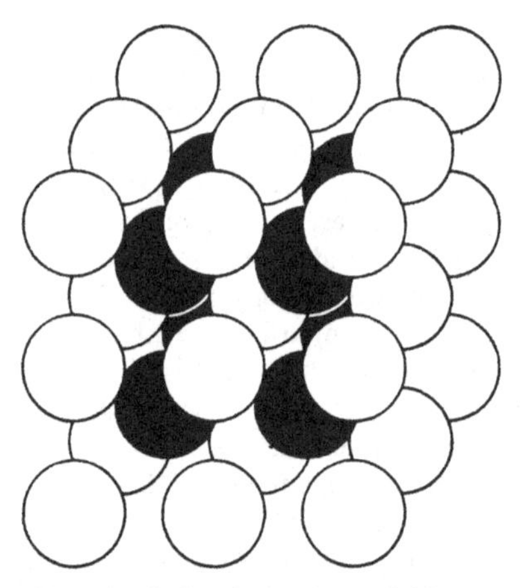

Fig. 49. Ordered structure of β-brass.

the actual intersection of the surfaces. Let us conclude our examination of higher-order transitions by inquiring into the extent to which any analogues of superheating and supercooling can be expected in the vicinity of the transition line. We may deduce immediately from Gorter's model and the ensuing discussion that superheating is inconceivable, since there is no continuation above the transition line of the surface corresponding to that phase which is stable at lower temperatures. We might also conclude that supercooling is a theoretical possibility, since the vapour in the vessel may remain in a supercooled, uncondensed state below T_0, but this conclusion will turn out to be misleading, since it is in this one, almost trivial, respect that Gorter's system is an unsound model of a real second-order transition.

The possibility of superheating and supercooling at a first-order transition is due to the essential difference in the constitution of the two phases concerned. Just as the volume contributions to their energies and entropies are different, so in general we must expect the surface contributions to differ also, so that there will be a surface tension at a boundary between the phases. This surface tension imposes what may be thought of, in mechanical terms, as a potential barrier between the two phases. If we imagine the condensation of a slightly supercooled vapour to proceed by the formation and subsequent growth of a drop of liquid, we may represent the availability of the whole system as a function of the radius of the drop, making the crude approximation that the contribution of the drop is simply its

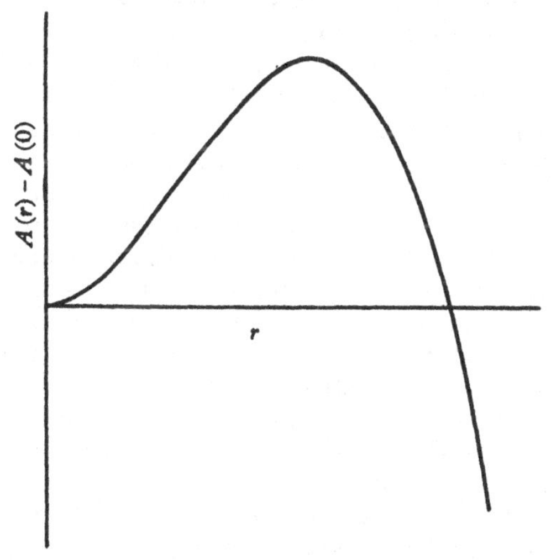

Fig. 50. Availability of vapour containing one liquid drop of radius r.

mass times the availability per unit mass of liquid, plus the surface tension times the area of the drop. If the vapour is supercooled, $g_v > g_l$, and the dependence of the availability on the radius of the drop is expressed by the equation

$$A(r) - A(0) = 4\pi r^2 \sigma - \tfrac{4}{3}\pi r^3 \rho_l (g_v - g_l),$$

in which $A(r)$ is the availability of the system containing a drop of radius r, $A(0)$ is the availability of the system before the drop is formed, σ is the surface tension and ρ_l the density of the liquid. As on p. 110, g_v and g_l must both be calculated at the pressure and temperature of the vapour. This expression is shown in fig. 50, from which it will be seen that although the availability is lowered by the production of a large drop, the early stages of formation of a drop involve a raising of A, which is precluded by the arguments of Chapter 7. This is in essence the reason why supercooling is possible, although the argument given here is too superficial to be entirely correct; taken at its face value it would imply that however great the supercooling there could never be any condensation to the liquid phase, except by the introduction of an extraneous impurity on which condensation could readily occur. A more sophisticated approach to the problem, by fluctuation theory or by treating the early stages of formation of the drop as a quasi-chemical reaction between gas molecules,† shows that there will always be present a very small number of droplets, and that condensation may in principle always occur when the vapour is supercooled, by the growth of these droplets. The supercooled vapour is thus not metastable in the strict mechanical sense that it corresponds to a local, rather than an absolute, minimum of A; it is rather to be regarded as slowly transforming itself into the liquid phase, but at so slow a rate as to be inappreciable. The rate may be calculated, and it is found that water vapour at room temperature and at a pressure which is twice the equilibrium vapour pressure should condense, without external aid, in about 10^{40} years. As the pressure is increased the time for condensation diminishes, until at six times the vapour pressure the time is calculated to be only a few seconds, and this accords fairly well with careful observations on pure water vapour. It is clear that to the purist the supercooled state is not acceptable as an equilibrium state. Nevertheless if the supersaturation is not too great, the contribution to the Gibbs function of the system by the small number of droplets present after a reasonable time is quite immeasurably small, so that the g-surface apparently runs smoothly across the equilibrium line. Under these circumstances, as discussed in Chapter 2, we can afford for thermodynamical purposes to treat the supercooled system as if it were genuinely metastable.

In Gorter's model the possibility of supercooling at the second-order transition arises from the possibility of supercooling of a vapour which, we have seen, is the consequence of a non-zero surface tension. But if we take account of surface tension in the model the transition ceases to be strictly of the second order. We may imagine the vessel just below the transition temperature to contain one droplet suspended in the vapour. On account of surface tension the equilibrium pressure in the vessel will be modified in accordance with Kelvin's vapour-pressure relation, (7·16), and will be higher than the vapour pressure over a plane surface. Now although (see p. 111) a drop surrounded by vapour maintained at constant pressure is at best in unstable equilibrium, if the vapour is confined to a fixed volume the equilibrium is stable for drops greater than a certain critical size; this is because the act of evaporating a little liquid from the drop increases the pressure of the vapour at a greater rate than is required by (7·16). But below

† J. Frenkel, *Kinetic Theory of Liquids* (Oxford, 1946), p. 382.

this critical size the drop is unstable and collapses suddenly. Thus in Gorter's model the introduction of surface tension implies that the amount of liquid does not fall smoothly to zero, and in fact the transition becomes one of the first order, albeit with a very minute latent heat. If we wish to simulate precisely a second-order transition we must put the surface tension equal to zero, and then we automatically eliminate the prime cause of supercooling in the vapour.

By this analysis we have removed any justification for taking the Gorter model as favouring the possibility of supercooling in a second-order transition, and indeed there is no reason to expect supercooling to occur. For, in contrast to a first-order transition, no meaning is attachable to the idea of a phase-boundary at a second-order transition, since the two 'phases' are identical along the transition line.† No discontinuity of properties occurs in crossing the transition line, and there is therefore nothing to inhibit the transition. In fact no supercooling has even been observed at any transition of higher than first order.‡ The behaviour of some superconductors exemplifies clearly the contrast between first- and second-order transitions. In a magnetic field the transition is of the first order and one may expect superheating and supercooling to be observed. Both phenomena have in fact been observed, and the latter is particularly striking in aluminium, which may when very pure be maintained in the normal (non-superconducting) state in a magnetic field whose strength is only one-twentieth of the critical field, $\mathscr{H}_c$. It has never, however, been found possible to reduce the field to zero at any temperature below T_c without the superconducting state being established. Diagrammatically the behaviour may be represented as in fig. 51; the central curve shows $\mathscr{H}_c$, the field at which $g'_n = g'_s$, while the upper and lower curves represent the limits of superheating and supercooling respectively. At any non-zero value of $\mathscr{H}$ both supercooling and, to a lesser extent, superheating are possible, but when $\mathscr{H} = 0$ the transition proceeds without hysteresis at the transition temperature T_c. At temperatures below T_c, with $\mathscr{H} = 0$, the normal state cannot exist in any sort of equilibrium, stable or metastable.

We must therefore conclude that from a thermodynamical point of

† This is true only for a homogeneous sample. In an inhomogeneous sample, as for instance a very tall cylinder in which, on account of gravity, the pressure varies with height, it is in principle possible to establish a phase boundary in equilibrium, and the stable position of the phase boundary will alter as the temperature is changed.

‡ It is often observed with alloys such as β-brass, which exhibit the order-disorder phenomenon, that the disordered state may be maintained at temperatures well below the transition temperature by sufficiently rapid cooling. The reason why this should not be regarded as analogous to supercooling will be explained on pp. 157–8.

view (and this conclusion is strengthened by a detailed analysis of real physical systems by statistical means) we are not justified in drawing a diagram like fig. 47*b* to represent a second-order transition. No physical meaning can in general be attached either to the line representing phase 1 above T_c or to the line representing phase 2 below T_c. It is far better to regard the *g*-surface as a single-valued function of *P* and *T*, which shows a discontinuity in its second derivatives along a certain line, the transition line. This provides the simplest, and physically most exact, picture of a second-order transition, and leads

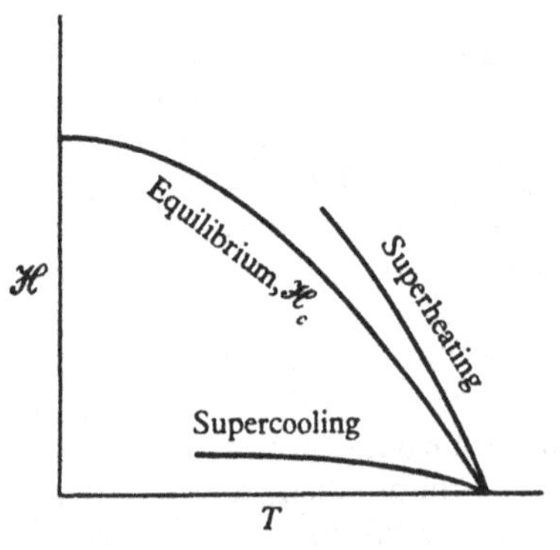

Fig. 51. Superheating and supercooling of a superconductor.

immediately to Ehrenfest's equations without admitting any validity to the criticisms which were based on analogy with a first-order transition. In no useful sense can the second-order transition be regarded as the limiting form of a first-order transition in which the latent heat has been allowed to tend to zero.

This conclusion is just as valid for other higher-order transitions, as may be illustrated by two simple examples. The perfect Bose-Einstein gas shows a third-order transition at its point of condensation (fig. 42). The theory of this phenomenon allows one and only one state of the gas for any given values of pressure and temperature, and reveals how the *g*-surface possesses a line of discontinuities in the third differential coefficients. There is no question here of any conceivable continuation of either branch of the surface across the transition line. As a second example let us consider a λ-transition such as that of β-brass. The theory of Bragg and Williams, mentioned above, led to a second-order transition at the point where the ordering process began as the temperature was lowered, but this is not true to the facts (fig. 41). In particular the rounded portion of the curve for C_P just above T_λ is evidence, as more refined theories show, that order is not completely destroyed at T_λ, but that there is a persistence of 'local order' to higher temperatures, in the sense that there remain in the equilibrium state small groups of atoms in which there is a marked tendency for alternation of copper and zinc atoms. If we choose any copper atom at random we shall find on the average that its nearest neighbours are rather more likely to be zinc that copper atoms, and the next nearest neighbours rather more likely to be copper than zinc. As we move away from the chosen atom, however, we shall find that this discrimination between lattice sites becomes less and less marked. It is possible to define the range of local order along the following lines. Starting from a given copper atom we may label alternate sites as C (copper) or Z (zinc) sites according to their occupants in a perfectly ordered arrangement. Then, selecting those sites which lie at a distance R from the chosen central copper atom we may count the number, r, of copper atoms on C-sites and zinc atoms on Z-sites, and the number, w, of copper atoms on Z-sites and zinc atoms on C-sites, and define the degree of order $\omega(R)$ by the equation

$$\omega(R) \equiv \frac{r-w}{r+w}.$$

In a completely ordered lattice $\omega = 1$, since $w = 0$; in a completely disordered lattice $\omega = 0$, since there is random occupation of the sites, and $r = w$. Let us now plot ω as a function of R; we may expect to find, after repeating this procedure many times for different central atoms, an average diagram such as fig. 52 and R_0, the width of the plot (to a value of R at which $\omega(R_0) = \frac{1}{2}\omega(0)$, say) is a measure of the range of local order. As the temperature is lowered towards T_λ, both $\omega(0)$ and R_0 increase until at T_λ the range of order becomes infinite; $\omega(R)$ does not tend to zero as R tends to infinity. Further lowering increases both $\omega(0)$ and the value to which $\omega(R)$ tends as R tends to infinity, which is a measure of the degree of long-range order. This picture of the transition fixes the transition point as the temperature at which long-range order first appears on cooling. But it will be seen that now there is no sharp distinction between disordered and partially ordered states; there is no definable state at temperatures above T_λ which can be extrapolated into metastable existence below T_λ without being in itself a state of long-range order.

In the preceding description of the ordering process it has been assumed that at every temperature there is an equilibrium state of order (either of short or long range), and that any experiment is performed sufficiently slowly that at each temperature the equilibrium state is established. It may now be argued that of necessity this precludes the occurrence of supercooling, just as a sufficiently slow cooling of a vapour through its condensation temperature would avoid the establishment of the supersaturated vapour phase. If the alloy is cooled rapidly from a temperature well above T_λ to one well below T_λ it may retain its disordered state virtually indefinitely on account of the very slow rate of migration of the atoms at low temperatures. There is, however, an important distinction to be made between these two cases. On the one hand a vapour just above its condensation temperature reaches equilibrium in a very short time, and on cooling to the supersaturated state just below the condensation temperature still retains this property of reaching what is apparently equilibrium extremely rapidly. As we have said before, the state is not strictly one of equilibrium since over a very long period condensation will occur. Still there is a sufficiently large margin between the short time for apparent equilibrium of the vapour (10^{-8} sec., say) and the long time of conden-

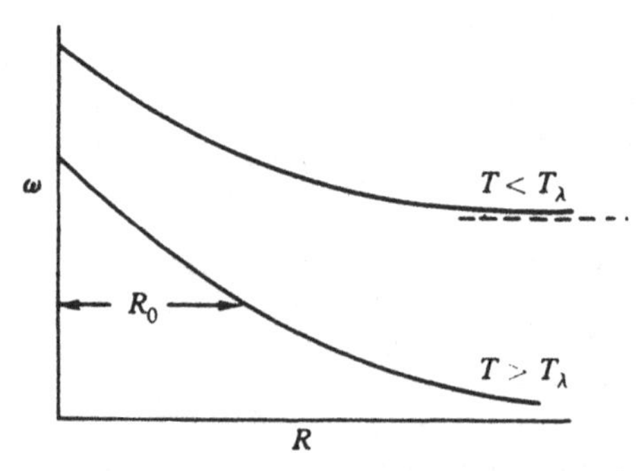

Fig. 52. Illustrating degree and range of local order.

sation (perhaps $> 10^{100}$ years for small supercooling of a pure vapour) that a meaning may be attached to the concept of a supercooled vapour in equilibrium. On the other hand the time taken for an alloy just above T_λ to reach its equilibrium state of partial short-range order is not significantly different from the time taken to establish long-range order just below T_λ, since the same migration mechanism is involved in both cases. When a disordered alloy is quenched by rapid cooling to a temperature well below T_λ, it is frozen into a particular configuration which is not analogous to the state of a supercooled vapour. For the state of the vapour is thermodynamically well defined, in that it does not depend on how fast the vapour was cooled or from what temperature, whereas the state of the quenched alloy shows more or less order according to the rate of cooling and the initial temperature. The only way in which the state of the whole alloy below T_λ may be made independent of the manner in which it was achieved is by allowing sufficient time for migration to occur, and this necessarily involves, as pointed out above, the production of the long-range ordered state, as being the only well-defined state at temperatures below T_λ. The existence of the quenching phenomenon does not therefore invalidate the view that true supercooling is never observed, and is indeed unobservable, in any but first-order transitions.

This last example has taken our reasoning well beyond the confines of classical thermodynamics, but the conclusion we arrive at is not dependent on the illustrations given of its application. The relations which may be derived for the thermodynamic behaviour in the vicinity of a transition line, of first or higher orders, in no case depend on the assumption that such a line marks the intersection or meeting of different surfaces; only for first-order transitions can such an assumption lead to a clarification rather than an actual misinterpretation of the physical situation.

EXERCISES

1. According to some textbooks, a knowledge of the Joule-Kelvin coefficient of a gas, and its specific heat at constant pressure, as functions of temperature and pressure is sufficient to enable the equation of state to be determined. Show that this is not so, and that further information, e.g. the form of one isotherm, is required.

2. A saturated vapour is expanded adiabatically; if L is the latent heat of vaporization, show that it becomes supersaturated or unsaturated according as whether $-T\frac{\mathrm{d}}{\mathrm{d}T}\left(\frac{L}{T}\right)$ is greater or less than the specific heat of the liquid under its own vapour pressure.

3. A simple liquefier is constructed so that compressed gas enters at room temperature T_0 and a high pressure P, and passes through a heat exchanger to a throttle, where it is expanded to a low pressure; part condenses and the rest returns through the heat exchanger, leaving the liquefier at room temperature and pressure. Show that the fraction of gas liquefied is greatest when the pressure of the gas entering is adjusted so that (P, T_0) is a point on the inversion curve of the gas.

4. A Simon helium liquefier consists essentially of a vessel into which helium gas is compressed to a high pressure P at 10° K. (above the critical point of helium). The vessel is then thermally isolated, and the gas is allowed to escape slowly through a capillary tube until the pressure within the vessel is 1 atmosphere, and the temperature 4·2° K., the normal boiling-point of helium. Assuming that the thermal isolation is perfect, that the heat capacity of the vessel is negligible in comparison with that of the gas, and that the gas obeys the perfect gas law, calculate what value of P must be chosen for the vessel to be entirely filled with liquid. [Latent heat of liquid helium at 4·2° K. = 20 cal. mole^{-1}; C_V of gaseous helium = 3 cal. mole^{-1} deg.$^{-1}$.]

5. Show that in a small Joule expansion of a fluid having negative expansion coefficient the pressure changes more than in the corresponding adiabatic expansion, the ratio of the two changes being $1-PV\beta/C_P$.

6. In a certain compressor gas at room temperature T_0 and atmospheric pressure P_0 is compressed adiabatically, and is then passed through water-cooled tubes until eventually it emerges at pressure P_1 and temperature T_0. Find an expression for the work required for this process, compared with what would be needed for a reversible isothermal compression leading to the same result, and

show that the ratio is not less than unity. Discuss also the changes of entropy occurring in the two processes.

7. An ellipsoid made of a magnetically isotropic substance is free to rotate about a vertical axis in a uniform horizontal magnetic field, with two unequal axes horizontal. The susceptibility is independent of field strength.

(*a*) Show both by direct calculation of couples and by the thermodynamic conditions for equilibrium that the ellipsoid tends to set itself with the longer horizontal axis (i.e. that having the smaller demagnetizing coefficient) parallel to the field, whether the substance is paramagnetic or diamagnetic.

(*b*) If the ellipsoid is thermally isolated, and constructed of a paramagnetic material which obeys Curie's law, derive an expression for the variation of its temperature when it is rotated in a constant field, in terms of the demagnetizing coefficients along the principal axes. For the sake of simplicity assume that the susceptibility is much smaller than unity.

8. For a system consisting of two phases of a substance in equilibrium, the specific heat at constant volume and the adiabatic compressibility are related by the equation

$$C_V/k_S = VT(\mathrm{d}P/\mathrm{d}T)^2,$$

where $\mathrm{d}P/\mathrm{d}T$ is the slope of the equilibrium line on the phase diagram; show that this result holds if the transition between the phases is either of the first order or a λ-transition.

9. On the basis of the following information, which is partly hypothesis and partly somewhat simplified experimental data, calculate the melting pressure of ${}_2\mathrm{He}^3$ at 0° K.:

(*a*) Between 0 and 10^{-5} ° K. the specific heat of the solid is very high, but between 10^{-5} and 1° K. it is much less than that of the liquid.

(*b*) The specific heat of the liquid is proportional to T below 1° K.

(*c*) The expansion coefficient of both phases may be assumed to be zero.

(*d*) At 0·32° K. the melting pressure P_m is 29·4 atmospheres and $\mathrm{d}P_m/\mathrm{d}T=0$; at 0·7° K. P_m is 33 atmospheres.

10. According to experimental measurements shown in fig. 42, α-iron transforms into γ-iron at 906° C. and back to α-iron at 1400° C. Between these temperatures the specific heat of γ-iron rises linearly from 0·160 cal. g.$^{-1}$ deg.$^{-1}$ to 0·169 cal. g.$^{-1}$ deg.$^{-1}$. On the assumption that α-iron, if it were stable between 906 and 1400° C., would have a specific heat constant at the value 0·185 cal. g.$^{-1}$ deg.$^{-1}$ that it has at both these temperatures, calculate the latent heat at each transition

and comment on the experimental value for the 906° C. transition, 3·86 cal. g.$^{-1}$.

11. The transition point of the $S^\alpha \to S^\beta$ transition is 95·5° C., and the melting point of S^β is 119·3° C., at atmospheric pressure. The latent heat of the transition $S^\alpha \to S^\beta$ is 2·78 cal. g.$^{-1}$, and the latent heat of fusion of S^β is 13·2 cal. g.$^{-1}$. The densities of S^α, S^β and liquid sulphur are 2·07, 1·96, and 1·90 g. cm.$^{-3}$, respectively. Assuming the latent heats and densities to be independent of temperature and pressure, find the co-ordinates (P, T) of the triple point of S^α, S^β, and liquid sulphur.

12. Discuss qualitatively the effects produced by gravity in experimental determinations of

(*a*) the specific heat of a substance exhibiting a second-order or λ-transition;

(*b*) the critical isotherm of a simple fluid.

13. Show that for a substance obeying van der Waals's equation, the latent heat drops to zero with a vertical tangent as the critical point is approached, i.e. $\lim_{T \to T_c} (\mathrm{d}L/\mathrm{d}T) = -\infty$. What behaviour is to be expected for a real substance such as xenon, for which the isotherms near T_c are shown in fig. 26?

14. The specific heat of unit volume of a metal may be very approximately represented by the formulae

$$C_s = aT^3 \quad \text{in the superconducting state,}$$
$$C_n = bT^3 + \gamma T \quad \text{in the normal state,}$$

where a, b and γ are constants.

Show that these formulae lead to the following results:

(*a*) the transition temperature in zero field, $T_c = (3\gamma/(a-b))^{\frac{1}{2}}$;

(*b*) the critical magnetic field $\mathscr{H}_c = \mathscr{H}_0(1-t^2)$ where $t = T/T_c$ and $\mathscr{H}_0 = T_c\sqrt{(2\pi\gamma)}$;

(*c*) the difference between the internal energies of the two states, in zero field, has a maximum when the temperature is $T_c/\sqrt{3}$;

(*d*) if a magnetic field applied to an isolated superconductor is increased very slowly to a value above the critical, the transition to the normal state is accompanied by a cooling of the metal. Find an expression for the drop in temperature;

(*e*) if the field is applied suddenly, instead of slowly, the metal is heated rather than cooled if the strength of the field exceeds

$$\mathscr{H}_0[(1+3t^2)(1-t^2)]^{\frac{1}{2}}.$$

Further exercises, some more straightforward than those given here, will be found in *Cavendish Problems in Classical Physics* (C.U.P. 1963).

INDEX